Progress in Soil Science

Aims and Scope

Progress in Soil Science series aims to publish books that contain novel approaches in soil science in its broadest sense – books should focus on true progress in a particular area of the soil science discipline. The scope of the series is to publish books that enhance the understanding of the functioning and diversity of soils in all parts of the globe. The series includes multidisciplinary approaches to soil studies and welcomes contributions of all soil science subdisciplines such as: soil genesis, geography and classification, soil chemistry, soil physics, soil biology, soil mineralogy, soil fertility and plant nutrition, soil and water conservation, pedometrics, digital soil mapping, proximal soil sensing, digital soil morphometrics, soils and land use change, global soil change, natural resources and the environment.

More information about this series at http://www.springer.com/series/8746

Brendan P. Malone • Budiman Minasny
Alex B. McBratney

Using R for Digital Soil Mapping

Springer

Brendan P. Malone
Sydney Institute of Agriculture
The University of Sydney
Eveleigh, NSW, Australia

Budiman Minasny
Sydney Institute of Agriculture
The University of Sydney
Eveleigh, NSW, Australia

Alex B. McBratney
Sydney Institute of Agriculture
The University of Sydney
Eveleigh, NSW, Australia

ISSN 2352-4774 ISSN 2352-4782 (electronic)
Progress in Soil Science
ISBN 978-3-319-44325-6 ISBN 978-3-319-44327-0 (eBook)
DOI 10.1007/978-3-319-44327-0

Library of Congress Control Number: 2016948860

Printed on acid-free paper

This Springer imprint is published by Springer Nature
The registered company is Springer International Publishing AG Switzerland

Foreword

Digital soil mapping is a runaway success. It has changed the way we approach soil resource assessment all over the world. New quantitative DSM products with associated uncertainty are appearing weekly. Many techniques and approaches have been developed. We can map the whole world or a farmer's field. All of this has happened since the turn of the millennium. DSM is now beginning to be taught in tertiary institutions everywhere. Government agencies and private companies are building capacity in this area. Both practitioners of conventional soil mapping methods and undergraduate and research students will benefit from following the easily laid out text and associated scripts in this book carefully crafted by Brendan Malone and colleagues. Have fun and welcome to the digital soil century.

Dominique Arrouays – Scientific coordinator of GlobalSoilMap.

Preface

Digital soil mapping (DSM) has evolved from a science-driven research phase of the early 1990s to presently a fully operational and functional process for spatial soil assessment and measurement. This evolution is evidenced by the increasing extents of DSM projects from small research areas towards regional, national and even continental extents.

Significant contributing factors to the evolution of DSM have been the advances in information technologies and computational efficiency in recent times. Such advances have motivated numerous initiatives around the world to build spatial data infrastructures aiming to facilitate the collection, maintenance, dissemination and use of spatial information. Essentially, fine-scaled earth resource information of improving qualities is gradually coming online. This is a boon for the advancement of DSM. More importantly, however, the contribution of the DSM community in general to the development of such generic spatial data infrastructure has been through the ongoing creation and population of regional, continental and worldwide soil databases from existing legacy soil information. Ambitious projects such as those proposed by the *GlobalSoilMap* consortium, whose objective is to generate a fine-scale 3D grid of a number of soil properties across the globe, provide some guide to where DSM is headed operationally. We are also seeing in some countries of the world the development of nationally consistent comprehensive digital soil information systems—the Australian Soil Grid http://www.clw.csiro.au/aclep/soilandlandscapegrid/ being particularly relevant in that regard. Besides the mapping of soil properties and classes, DSM approaches have been extended to other soil spatial analysis domains such as those of digital soil assessment (DSA) and digital soil risk assessment (DSRA).

It is an exciting time to be involved in DSM. But with development and an increase in the operational status of DSM, there comes a requirement to teach, share and spread the knowledge of DSM. Put more simply, there is a need to teach more people how to do it. It is such that this book attempts to share and disseminate some of that knowledge.

The focus of the materials contained in the book is to learn how to carry out DSM in a real work situation. It is procedural and attempts to give the participant a taste and a conceptual framework to undertake DSM in their own technical fields. The book is very instructional—a manual of sorts—and therefore completely interactive in that participants can access and use the available data and complete exercises using the available computer scripts. The examples and exercises in the book are delivered using the `R` computer programming environment. Subsequently, this course is both training in DSM *and* `R`. Using `R`, this course will introduce some basic `R` operations and functionality in order to gain some fluency in this popular scripting language. The DSM exercises will cover procedures for handling and manipulating soil and spatial data in `R` and then introduce some basic concepts and practices relevant to DSM, which importantly includes the creation of digital soil maps. As you will discover, DSM is a broad term that entails many applications, of which a few are covered in this book.

The material contained in this book has been cobbled together over successive years from 2009. This effort has largely been motivated by the need to prepare a hands-on DSM training course with associated materials as an outreach programme of the Pedometrics and Soil Security research group at the University of Sydney. The various DSM workshops have been delivered to a diverse range of participants: from undergraduates, to postgraduates, to tenured academics, as well as both private and government scientists and consultants. These workshops have been held both at the Soil Security laboratories at the University of Sydney, as well as various locations around the world. The ongoing development of teaching materials for DSM needs to continue over time as new discoveries and efficiencies are made in the field of DSM and, more generally, pedometrics. Therefore, we would be very grateful to receive feedback and suggestions on ways to improve the book so that the materials remain accessible, up to date and relevant.

Eveleigh, Australia

Brendan P. Malone
Budiman Minasny
Alex B. McBratney

Endorsements

This book entitled *Using R for Digital Soil Mapping* is an excellent book that clearly outlines the step-by-step procedures required for many aspects of digital soil mapping. This is my first time to learn R language and spatial modelling for DSM, but with the instructive book, it's easy to produce different DSMs by following text and associate R scripts. It has been especially useful in Taiwan for soil organic carbon stock mapping in different soil depths and of different parent materials and different land uses. The other good experience is the clear pointers on how to prepare the covariates to build the spatial prediction functions for DSM by regression models if we do not have enough soil data. I strongly recommend this excellent book to any person to apply DSM techniques for studying the spatial variability of agriculture and environmental sciences.

Distinguished Professor Zueng-Sang Chen, Department of Agricultural Chemistry, National Taiwan University, Taipei, Taiwan.

I can recommend this book as an excellent support for those wanting to learn digital soil mapping methods. The hands-on exercises provide invaluable examples of code for implementing in the R computing language. The book will certainly assist you to develop skills in R. It will also introduce you to a very wide range of powerful numerical and categorical modelling approaches that are emerging to enable quantitative spatial and temporal inference of soil attributes at all scales from local to global. There is also a valuable chapter on how to assess uncertainty of the digital soil map that has been produced. The book exemplifies the quantum leap that is occurring in quantitative spatial and temporal modelling of soil attributes, and is a must for students of this discipline.

Carolyn Hedley, Soil Scientist, New Zealand.

Using R for Digital Soil Mapping is a fantastic resource that has enabled us to develop and build our skills in digital soil mapping (DSM) from scratch, so much so that this discipline has now become part of our agency core business in Tasmanian land evaluation. It's thorough instructional content has enabled us to deliver a state-wide agricultural enterprise suitability mapping programme, developing quantitative soil property surfaces with uncertainties through predictive spatial modelling, including covariate processing, optimised soil sampling strategies and standardised soil depth-spline functions. We continually refer to this 'easy to follow' guide when developing the necessary R-code to undertake our DSM; using the freely available R environment rather than commercial software in itself has saved thousands of dollars in software fees and allowed automation and time-saving in many DSM tasks. This book is a must for any individual, academic institution or government soil agency wishing to embark into the rapidly developing world of DSM for land evaluation, and will definitely ease the 'steepness' in the learning curve.

Darren Kidd, Department of Primary Industries Parks Water and Environment, Tasmania, Australia.

This excellent book contains clear step-by-step examples in digital soil mapping (DSM), such as how to prepare covariates, to build spatial prediction functions using either regression or classification models and to apply the prediction functions to produce maps and their uncertainties. When I started my research in DSM, I have very little experience in R and spatial modelling. By following clear instructions presented in this book, I have succeeded in learning and developing DSM techniques for mapping the depth and carbon stock in Indonesian tropical peatlands. I highly recommend this book to anyone who wants to learn and apply DSM techniques.

Rudiyanto, Institut Pertanian Bogor, Indonesia.

Acknowledgements

Special thanks to those who have contributed to the development of materials in this book. Pierre Roudier is pretty much solely responsible for helping put together the materials regarding interactive mapping and the caret package for digital soil mapping. Colleagues at the University of Sydney, especially Uta Stockmann, have given continual feedback throughout the development of the DSM teaching materials of the past number of years. Lastly, we are grateful to the numerous participants of our DSM workshops throughout the world. With their feedback and questions, the materials have evolved and been honed over time to make this a reasonably substantial one-stop shop for practicable DSM. Cheers to all!

Contents

Chapter 1
Digital Soil Mapping

1.1 The Fundamentals of Digital Soil Mapping

In recent times we have bared witness to the advancement of the computer and information technology ages. With such advances, there have come vast amounts of data and tools in all fields of endeavor. This has motivated numerous initiatives around the world to build spatial data infrastructures aiming to facilitate the collection, maintenance, dissemination and use of spatial information. Soil science potentially contributes to the development of such generic spatial data infrastructure through the ongoing creation of regional, continental and worldwide soil databases, and which are now operational for some uses e.g., land resource assessment and risk evaluation (Lagacherie and McBratney 2006).

Unfortunately the existing soil databases are neither exhaustive enough nor precise enough for promoting an extensive and credible use of the soil information within the spatial data infrastructure that is being developed worldwide. The main reason is that their present capacities only allow the storage of data from conventional soil surveys which are scarce and sporadically available (Lagacherie and McBratney 2006).

The main reason for this lack of soil spatial data is simply that conventional soil survey methods are relatively slow and expensive. Furthermore, we have also witnessed a global reduction in soil science funding that started in the 1980s (Hartemink and McBratney 2008), which has meant a significant scaling back in wide scale soil spatial data collection and/or conventional soil surveying.

To face this situation, it is necessary for the current spatial soil information systems to extend their functionality from the storage and use of digitized (existing) soil maps, to the production of soil maps ab initio (Lagacherie and McBratney 2006). This is precisely the aim of Digital Soil Mapping (DSM) which can be defined as:

B.P. Malone et al., *Using R for Digital Soil Mapping*,
Progress in Soil Science, DOI 10.1007/978-3-319-44327-0_1

> The creation and population of spatial soil information systems by numerical models inferring the spatial and temporal variations of soil types and soil properties from soil observation and knowledge from related environmental variables. (Lagacherie and McBratney 2006)

The concepts and methodologies for DSM were formalized in an extensive review by McBratney et al. (2003). In the McBratney et al. (2003) paper, the *scorpan* approach for predictive modelling (and mapping) of soil was introduced, which in itself is rooted in earlier works by Jenny (1941) and Russian soil scientist Dokuchaev. *scorpan* is a mnemonic for factors for prediction of soil attributes: soil, climate, organisms, relief, parent materials, age, and spatial position. The *scorpan* approach is formulated by the equation:

$$S = f(s, o, r, r, p, a, n) + \epsilon$$

or

$$S = f(Q) + \epsilon$$

Long-handed, the equation states that the soil type or attribute at an unvisited site *(S)* can be predicted from a numerical function or model *(f)* given the factors just described plus the locally varying, spatial dependent residuals (ϵ). The *f(Q)* part of the formulation is the deterministic component or in other words, the empirical quantitative function linking S to the *scorpan* factors (Lagacherie and McBratney 2006). The *scorpan* factors or environmental covariates come in the form of spatially populated digitally available data, for instance from digital elevation models and the indices derived from them—slope, aspect, MRVBF etc. Landsat data, and other remote sensing images, radiometric data, geological survey maps, legacy soil maps and data, just to name a few. For the residuals (ϵ) part of the formulation, we assume there to be some spatial structure. This is for a number of reasons which include that the attributes used in the deterministic component were inadequate, interactions between attributes were not taken into account, or the form of *f()* was mis-specified. Overall this general formulation is called the *scorpan* kriging method, where the kriging component is the process of defining the spatial trend of the residuals (with variograms) and using kriging to estimate the residuals at the non-visited sites.

Without getting into detail with regards to some of the statistical nuances such as bias issues—which can be prevalent when using legacy soil point data for DSM—that are encountered with using this type of data, the application of *scorpan* kriging can only be done in extents where there is available soil point data. The challenge therefore is: what to do in situations where this type of data is not available? In the context of the global soil mapping key soil attributes, this is a problem, but can be overcome with the usage of other sources of legacy soil data such as existing soil maps. It is even more of a problem when this information is not available either. However, in the context of global soil mapping, Minasny and McBratney (2010) proposed a decision tree structure for actioning DSM on the basis of the nature of available legacy soil data. This is summarized in Fig. 1.1. But bear in mind that this

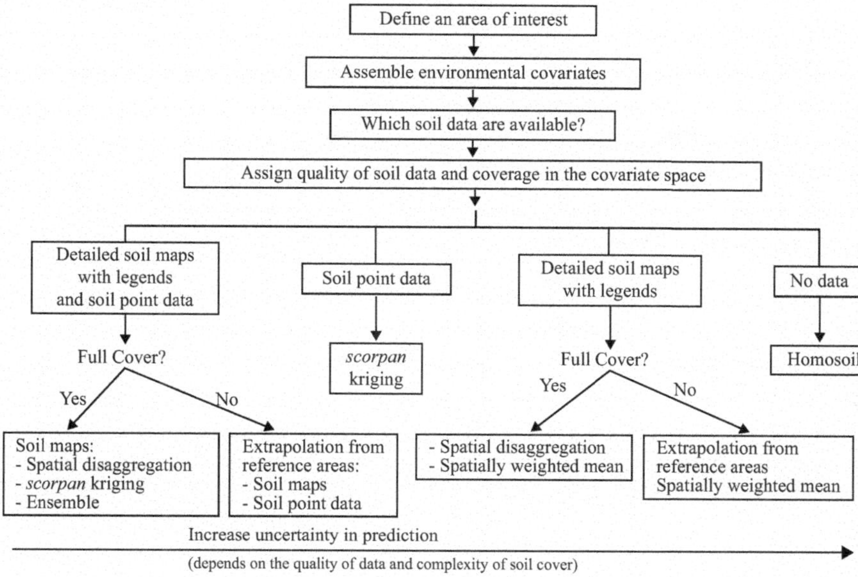

Fig. 1.1 A decision tree for digital soil mapping based on legacy soil data (Adapted from Minasny and McBratney 2010)

decision tree is not constrained only to DSM at a global scale but at any mapping extent where the user wishes to perform DSM given the availability of soil data for their particular area.

As can be seen from Fig. 1.1, once you have defined an area of interest, and assembled a suite of environmental covariates for that area, then determined the availability of the soil data there, you follow the respective pathway. *scorpan* kriging is performed exclusively when there is only point data, but can be used also when there is both point and map data available, e.g., (Malone et al. 2014). The work flow is quite different when there is only soil map information available. Bear in mind that the quality of the soil map depends on the scale and subsequently variation of soil cover; such that smaller scaled maps e.g., 1:100,000 would be considered better and more detailed than large scaled maps e.g., 1:500,000. The elemental basis for extracting soil properties from legacy soil maps comes from the central and distributional concepts of soil mapping units. For example, modal soil profile data of soil classes can be used to quickly build soil property maps. Where mapping units consist of more than one component, we can use a spatially weighted means type method i.e., estimation of the soil properties is based on the modal profile of the components and the proportional area of the mapping unit each component covers, e.g., (Odgers et al. 2012). As a pre-processing step prior to creating soil attribute maps, it may be necessary to harmonize soil mapping units (in the case of adjacent soil maps) and/or perform some type of disaggregation technique in order to retrieve the map unit component information. Some approaches for doing so have been described in Bui and Moran (2003). More recently soil map disaggregation has been a target of DSM interest with a sound contribution from Odgers et al. (2014) for extracting individual soil series or soil class information from convolved soil map units by way of the DSMART algorithm. The DSMART algorithm can best be explained as a data mining with repeated re-sampling algorithm. Furthering the DSMART algorithm, Odgers et al. (2015) then introduced the PROPR algorithm which takes probability outputs from DSMART together with modal soil profile data of given soil classes, to estimate soil attribute quantities (with estimates of uncertainty).

What is the process when there is no soil data available at all? This is obviously quite a difficult situation to confront, but a real one at that. The central concept that was discussed by Minasny and McBratney (2010) for addressing these situations is based on the assumed homology of soil forming factors between a reference area and the region of interest for mapping. Malone et al. (2016) provides a further overview of the topic together with a real world application which compared different extrapolating functions. Overall, the soil homologue concept or Homosoil, relative to other areas of DSM research is still in its development. But considering from a global perspective, the sparseness of soil data and limited research funds for new soil survey, application of the Homosoil approach or other analogues will become increasingly important for the operational advancement of DSM.

1.2 What Is Going to Be Covered in this Book?

This book covers some of the territory that is described in Fig. 1.1, particularly the *scorpan* kriging type approach of DSM; as this is probably most commonly undertaken. Also covered is spatial disagregation of polygonal maps. This is framed in the context of updating digital soil maps and downscaling in terms of deriving soil class or attribute information from aggregated soil mapping units. Importantly there is a theme of implementation about this book; a sort of how to guide. So there are some examples of how to create digital soil maps of both continuous and categorical target variable data, given available points and a portfolio of available covariates. The procedural detail is explained and implemented using the `R` computing language. Subsequently, some effort is required to become literate in this programming language, both for general purpose usage and for DSM and other related soil studies. With a few exceptions, all the data that is used in this book to demonstrate methods, together with additional functions are provided via the R package: *ithir*. This package can be downloaded free of cost. Instructions for getting this package are in the next chapter.

The motivation of the book then shifts to operational concerns and based around real case-studies. For example, the book looks at how we might statistically validate a digital soil map. Another operational study is that of digital soil assessment (Carre et al. 2007). Digital soil assessment (DSA) is akin to the translation of digital soil maps into decision making aids. These could be risk-based assessments, or assessing threats to soil (erosion, decline of organic matter etc.), and assessing soil functions. These type of assessments can not be easily derived from a digital soil map alone, but require some form of post-processing inference. This could be done with quantitative modeling and or a deep mechanistic understanding of the assessment that needs to be made. A natural candidate in this realm of DSM is land capability or agricultural enterprise suitability. A case study of this type of DSA is demonstrated in this book. Specific topics of this book include:

1. Attainment of `R` literacy in general and for DSM.
2. Algorithmic development for soil science.
3. General GIS operations relevant to DSM.
4. Soil data preparation, examination and harmonization for DSM.
5. Quantitative functions for continuous and categorical (and combinations of both) soil attribute modeling and mapping.
6. Quantifying digital soil map uncertainty.
7. Assessing the quality of digital soil maps.
8. Updating, harmonizing and disaggregating legacy soil mapping.
9. Digital soil assessment in terms of land suitability for agricultural enterprises.
10. Digital identification of soil homologues.

References

Bui E, Moran CJ (2003) A strategy to fill gaps in soil survey over large spatial extents: an example from the Murray-Darling basin of Australia. Geoderma 111:21–41

Carre F, McBratney AB, Mayr T, Montanarella L (2007) Digital soil assessments: beyond DSM. Geoderma 142(1–2):69–79

Hartemink AE, McBratney AB (2008) A soil science renaissance. Geoderma 148:123–129

Jenny H (1941) Factors of soil formation. McGraw-Hill, New York

Lagacherie P, McBratney AB (2006) Digital soil mapping: an introductory perspective, chapter 1. In: Spatial soil information systems and spatial soil inference systems: perspectives for digital soil mapping. Elsevier, Amsterdam, pp 3–22

Malone BP, Minasny B, Odgers NP, McBratney AB (2014) Using model averaging to combine soil property rasters from legacy soil maps and from point data. Geoderma 232–234:34–44

Malone BP, Jha SK, Minasny AB, McBratney B (2016) Comparing regression-based digital soil mapping and multiple-point geostatistics for the spatial extrapolation of soil data. Geoderma 262:243–253

McBratney AB, Mendonca Santos ML, Minasny B (2003) On digital soil mapping. Geoderma 117:3–52

Minasny B, McBratney AB (2010) Digital soil mapping: bridging research, environmental application, and operation, chapter 34. In: Methodologies for global soil mapping. Springer, Dordrecht, pp 429–425

Odgers NP, Libohova Z, Thompson JA (2012) Equal-area spline functions applied to a legacy soil database to create weighted-means maps of soil organic carbon at a continental scale. Georderma 189–190:153–163

Odgers NP, McBratney AB, Minasny B (2015) Digital soil property mapping and uncertainty estimation using soil class probability rasters. Geoderma 237–238:190–198

Odgers NP, Sun W, McBratney AB, Minasny B, Clifford D (2014) Disaggregating and harmonising soil map units through resampled classification trees. Geoderma 214–215:91–100

Chapter 2
R Literacy for Digital Soil Mapping

2.1 Objective

The immediate objective here is to skill up in data analytics and basic graphics with R. The range of analysis that can be completed, and the types of graphics that can be created in R is simply astounding. In addition to the wide variety of functions available in the "base" packages that are installed with R, more than 4500 contributed packages are available for download, each with its own suite of functions. Some individual packages are the subject of entire books.

For this chapter of the book and the later chapters that will deal with digital soil mapping exercises, we will not be able to cover every type of analysis or plot that R can be used for, or even every subtlety associated with each function covered in this entire book. Given it's inherent flexibility, R is difficult to master, as one may be able to do with a stand-alone software. R is a software package one can only increase their knowledge and fluency in. Meaning that, effectively, learning R is a boundless pursuit of knowledge.

In a disclaimer of sorts, this introduction to R borrows many ideas, and structures from the plethora of online materials that are freely available on the internet. It will be worth your while to do a Google search from time-to-time if you get stuck—you will be amazed to find how many other R users have had the same problems you have or have had.

2.2 Introduction to R

2.2.1 R Overview and History

R is a software system for computations and graphics. According to the R FAQ (http://cran.r-project.org/doc/FAQ/R-FAQ.html#R-Basics):

B.P. Malone et al., *Using R for Digital Soil Mapping*,
Progress in Soil Science, DOI 10.1007/978-3-319-44327-0_2

It consists of a language plus a run-time environment with graphics, a debugger, access to certain system functions, and the ability to run programs stored in script files.

R was originally developed in 1992 by Ross Ihaka and Robert Gentleman at the University of Auckland (New Zealand). The R language is a dialect of the S language which was developed by John Chambers at Bell Laboratories. This software is currently maintained by the R Development Core Team, which consists of more than a dozen people, and includes Ihaka, Gentleman, and Chambers. Additionally, many other people have contributed code to R since it was first released. The source code for R is available under the GNU General Public Licence, meaning that users can modify, copy, and redistribute the software or derivatives, as long as the modified source code is made available. R is regularly updated, however, changes are usually not major.

2.2.2 Finding and Installing R

R is available for Windows, Mac, and Linux operating systems. Installation files and instructions can be downloaded from the Comprehensive R Archive Network (CRAN) site at http://cran.r-project.org/. Although the graphical user interface (GUI) differs slightly across systems, the R commands do not.

2.2.3 Running R: GUI and Scripts

There are two basic ways to use R on your machine: through the GUI, where R evaluates your code and returns results as you work, or by writing, saving, and then running R script files. R script files (or scripts) are just text files that contain the same types of R commands that you can submit to the GUI. Scripts can be submitted to R using the Windows command prompt, other shells, batch files, or the R GUI. All the code covered in this book is or is able to be saved in a script file, which then can be submitted to R. Working directly in the R GUI is great for the early stages of code development, where much experimentation and trial-and-error occurs. For any code that you want to save, rerun, and modify, you should consider working with R scripts.

So, how do you work with scripts? Any simple text editor works—you just need to save text in the ASCII format i.e., "unformatted" text. You can save your scripts and either call them up using the command `source ("file_name.R")` in the R GUI, or, if you are using a shell (e.g., Windows command prompt) then type `R CMD BATCH file_name.R`. The Windows and Mac versions of the R GUI comes with a basic script editor, shown below in Fig. 2.1.

Unfortunately, this editor is not very good by reason that the Windows version does not have syntax highlighting.

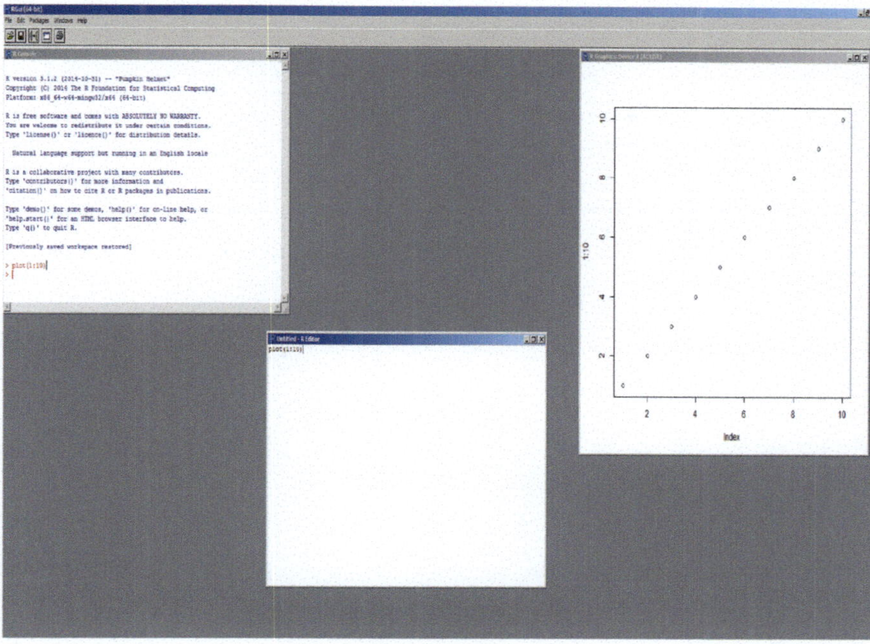

Fig. 2.1 R GUI, its basic script editor, and plot window

There are some useful (in most cases, free) text editors available that can be set up with R syntax highlighting and other features. TINN-R is a free text editor http://nbcgib.uesc.br/lec/software/des/editores/tinn-r/en that is designed specifically for working with R script files. Notepad++ is a general purpose text editor, but includes syntax highlighting and the ability to send code directly to R with the NppToR plugin. A list of text editors that work well with R can be found at: http://www.sciviews.org/_rgui/projects/Editors.html.

2.2.4 RStudio

RStudio http://www.rstudio.com/ is an integrated development environment (IDE) for R that runs on Linux, Windows and Mac OS X. We will be using this IDE during the book, generally because it is very well designed, intuitively organized, and quite stable.

When you first launch RStudio, you will be greeted by an interface that will look similar to that in Fig. 2.2.

The frame on the upper right contains the *workspace* (where you will be able see all your R objects), as well of a history of the commands that you have previously entered. Any plots that you generate will show up in the region in the lower right

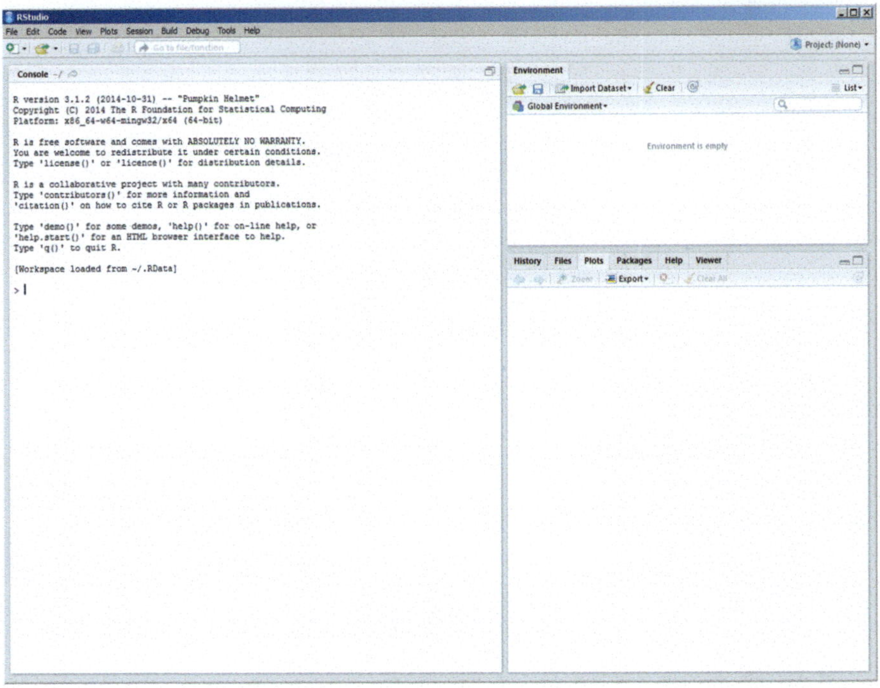

Fig. 2.2 The RStudio IDE

corner. Also in this region is various help documentation, plus information and documentation regarding what packages and function are currently available to use.

The frame on the left is where the action happens. This is the `R` console. Every time you launch RStudio, it will have the same text at the top of the console telling you the version that is being used. Below that information is the prompt. As the name suggests, this is where you enter commands into `R`. So lets enter some commands.

2.2.5 *R Basics: Commands, Expressions, Assignments, Operators, Objects*

Before we start anything, it is good to get into the habit of making scripts of our work. With RStudio launched go the *File* menu, then *new*, and *R Script*. A new blank window will open on the top left panel. Here you can enter your `R` prompts. For example, type the following: `1+1`. Now roll your pointer over the top of the panel to the right pointing green arrow (first one), which is a button for running the line of code down to the R console. Click this button and R will evaluate it. In the console you should see something like the following:

```
1 + 1

## [1] 2
```

You could have just entered the command directly into the prompt and gotten the same result. Try it now for yourself. You will notice a couple of things about this code. The > character is the prompt that will always be present in the GUI. The line following the command starts with a `[1]`, which is simply the position of the adjacent element in the output—this will make some sense later.

For the above command, the result is printed to the screen and lost—there is no assignment involved. In order to do anything other than the simplest analyses, you must be able to store and recall data. In R, you can assign the results of commands to symbolic variables (as in other computer languages) using the assignment operator `<-`. Note that other computer scripting languages often use the equals sign (=) as the assignment operator. When a command is used for assignment, the result is no longer printed to the GUI console.

```
x <- 1 + 1
x

## [1] 2
```

Note that this is very different from:

```
x < -1 + 1

## [1] FALSE
```

In this case, putting a space between the two characters that make up the assignment operator causes R to interpret the command as an expression that ask if `x` is less than zero. However spaces usually do not matter in R, as long as they do not separate a single operator or a variable name. This, for example, is fine:

```
x <- 1
```

Note that you can recall a previous command in the R GUI by hitting the up arrow on your keyboard. This becomes handy when you are debugging code.

When you give R an assignment, such as the one above, the object referred to as `x` is stored into the R workspace. You can see what is stored in the workspace by looking to the workspace panel in RStudio (top right panel). Alternatively, you can use the `ls` function.

```
ls()

## [1] "x"
```

To remove objects from your workspace, use `rm`.

```
rm(x)
x
```

As you can see, You will get an error if you try to evaluate what `x` is.

If you want to assign the same value to several symbolic variables, you can use the following syntax.

```
x <- y <- z <- 1
ls()

## [1] "x" "y" "z"
```

R is a case-sensitive language. This is true for symbolic variable names, function names, and everything else in R.

```
x <- 1 + 1
x
X
```

In R, commands can be separated by moving onto a new line (i.e., hitting enter) or by typing a semicolon (;), which can be handy in scripts for condensing code. If a command is not completed in one line (by design or error), the typical R prompt > is replaced with a +.

```
x<-
+  1+1
```

There are several operators that are used in the R language. Some of the more common are listed below.

Arithmetic

`+ - * / ^`	plus, minus, multiply, divide, power

Relational

`a == b`	a is equal to b (do not confuse with =)
`a != b`	a is not equal to b
`a < b`	a is less than b
`a > b`	a is greater than b
`a <= b`	a is less than or equal to b
`a >= b`	a is greater than or equal to b

Logical/grouping

!	not
&	and
|	or

Indexing

`$`	part of a data frame
`[]`	part of a data frame, array, list
`[[]]`	part of a list

Grouping commands

`{}`	specifying a function, for loop, if statement etc.

Making sequences

`a:b`	returns the sequence a, a+1, a+2, ... b

Others

`#`	commenting (very very useful!)
`;`	alternative for separating commands
`~`	model formula specification
`()`	order of operations, function arguments

Commands in R operate on objects, which can be thought of as anything that can be assigned to a symbolic variable. Objects include vectors, matrices, factors, lists, data frames, and functions. Excluding functions, these objects are also referred to as data structures or data objects.

When you want to finish up on an R session, RSudio will ask you if you want to "save workspace image". This refers to the workspace that you have created , i.e., all the objects you have created or even loaded. It is generally good practice to save your workspace after each session. More importantly however, is the need to save all the commands that you have created on your script file. Saving a script file in Rstudio is just like saving a Word document. Give both a go—save the script file and then save the workspace. You can then close RStudio.

2.2.6 R Data Types

The term "data type" refers to the type of data that is present in a data structure, and does not describe the data structure itself. There are four common types of data in R: numerical, character, logical, and complex numbers. These are referred to as *modes* and are shown below:
Numerical data

```
x <- 10.2
x
```

```
## [1] 10.2
```

Character data

```
name <- "John Doe"
name
```

```
## [1] "John Doe"
```

Any time character data are entered in the R GUI, you must surround individual elements with quotes. Otherwise, R will look for an object.

```
name <- John
```

```
## Error in eval(expr, envir, enclos): object 'John' not found
```

Either single or double quotes can be used in R. When character data are read into R from a file, the quotes are not necessary.

Logical data contain only three values: TRUE, FALSE, or NA, (NA indicates a missing value—more on this later). R will also recognize T and F, (for true and false respectively), but these are not reserved, and can therefore be overwritten by the user, and it is therefore good to avoid using these shortened terms.

```
a <- TRUE
a
```

```
## [1] TRUE
```

Note that there are no quotes around the logical values—this would make them character data. R will return logical data for any relational expression submitted to it.

```
4 < 2
```

```
## [1] FALSE
```

or

```
b <- 4 < 2
b
```

```
## [1] FALSE
```

And finally, complex numbers, which will not be covered in this book, are the final data type in R

```
cnum1 <- 10 + (0+3i)
cnum1
```

```
## [1] 10+3i
```

You can use the mode or class function to see what type of data is stored in any symbolic variable.

```
class(name)

## [1] "character"

class(a)

## [1] "logical"

class(x)

## [1] "numeric"

mode(x)

## [1] "numeric"
```

2.2.7 R Data Structures

Data in R are stored in data structures (also known as data objects)—these are and will be the that you perform calculations on, plot data from, etc. Data structures in R include vectors, matrices, arrays, data frames, lists, and factors. In a following section we will learn how to make use of these different data structures. The examples below simply give you an idea of their structure.

Vectors are perhaps the most important type of data structure in R. A vector is simply an ordered collection of elements (e.g., individual numbers).

```
x <- 1:12
x

##  [1]  1  2  3  4  5  6  7  8  9 10 11 12
```

Matrices are similar to vectors, but have two dimensions.

```
X <- matrix(1:12, nrow = 3)
X

##      [,1] [,2] [,3] [,4]
## [1,]    1    4    7   10
## [2,]    2    5    8   11
## [3,]    3    6    9   12
```

Arrays are similar to matrices, but can have more than two dimensions.

```
Y <- array(1:30, dim = c(2, 5, 3))
Y

## , , 1
##
##      [,1] [,2] [,3] [,4] [,5]
```

```
## [1,]    1    3    5    7    9
## [2,]    2    4    6    8   10
##
## , , 2
##
##      [,1] [,2] [,3] [,4] [,5]
## [1,]   11   13   15   17   19
## [2,]   12   14   16   18   20
##
## , , 3
##
##      [,1] [,2] [,3] [,4] [,5]
## [1,]   21   23   25   27   29
## [2,]   22   24   26   28   30
```

One feature that is shared for vectors, matrices, and arrays is that they can only store one type of data at once, be it numerical, character, or logical. Technically speaking, these data structures can only contain elements of the same mode.

Data frames are similar to matrices—they are two-dimensional. However, a data frame can contain columns with different modes. Data frames are similar to data sets used in other statistical programs: each column represents some variable, and each row usually represents an "observation", or "record", or "experimental unit".

```
dat <- (data.frame(profile_id = c("Chromosol", "Vertosol", "Sodosol"),
FID = c("a1", "a10", "a11"), easting = c(337859, 344059, 347034),
northing = c(6372415, 6376715, 6372740), visited = c(TRUE, FALSE, TRUE)))

dat
```

```
##   profile_id FID easting northing visited
## 1  Chromosol  a1  337859  6372415    TRUE
## 2   Vertosol a10  344059  6376715   FALSE
## 3    Sodosol a11  347034  6372740    TRUE
```

Lists are similar to vectors, in that they are an ordered collection of elements, but with lists, the elements can be other data objects (the elements can even be other lists). Lists are important in the output from many different functions. In the code below, the variables defined above are used to form a list.

```
summary.1 <- list(1.2, x, Y, dat)
summary.1
```

```
## [[1]]
## [1] 1.2
##
## [[2]]
##  [1]  1  2  3  4  5  6  7  8  9 10 11 12
##
## [[3]]
## , , 1
##
```

```
##      [,1] [,2] [,3] [,4] [,5]
## [1,]    1    3    5    7    9
## [2,]    2    4    6    8   10
##
## , , 2
##
##      [,1] [,2] [,3] [,4] [,5]
## [1,]   11   13   15   17   19
## [2,]   12   14   16   18   20
##
## , , 3
##
##      [,1] [,2] [,3] [,4] [,5]
## [1,]   21   23   25   27   29
## [2,]   22   24   26   28   30
##
##
## [[4]]
##   profile_id FID easting northing visited
## 1  Chromosol  a1  337859  6372415    TRUE
## 2   Vertosol a10  344059  6376715   FALSE
## 3    Sodosol a11  347034  6372740    TRUE
```

Note that a particular data structure need not contain data to exist. This may seem unusual, but it can be useful when it is necessary to set up an object for holding some data later on.

```
x <- NULL
```

2.2.8 *Missing, Indefinite, and Infinite Values*

Real data sets often contain missing values. R uses the marker NA (for "not available") to indicate a missing value. Any operation carried out on an NA will return NA.

```
x <- NA
x - 2
```

```
## [1] NA
```

Note that the NA used in R does not have the quotes around it—this would make it character data. To determine if a value is missing, use the `is.na`—this function can also be used to set elements in a data object to NA.

```
is.na(x)
```

```
## [1] TRUE
```

```
!is.na(x)

## [1] FALSE
```

Indefinite values are indicated with the marker `NaN`, for “not a number”. Infinite values are indicated with the markers `Inf` or `-Inf`. You can find these values with the functions `is.infinite, is.finite`, and `is.nan`.

2.2.9 Functions, Arguments, and Packages

In `R`, you can carry out complicated and tedious procedures using functions. Functions require arguments, which include the object(s) that the function should act upon. For example, the function `sum` will calculate the sum of all of its arguments.

```
sum(1, 12.5, 3.33, 5, 88)

## [1] 109.83
```

The arguments in (most) `R` functions can be named, i.e., by typing the name of the argument, an equal sign, and the argument value (arguments specified in this way are also called tagged). For example, for the function `plot`, the help file lists the following arguments.

```
plot (x, y,...)
```

Therefore, we can call up this function with the following code.

```
a <- 1:10
b <- a
plot(x = a, y = b)
```

With names arguments, `R` recognizes the argument keyword (e.g., `x` or `y`) and assigns the given object (e.g., `a` or `b` above) to the correct argument. When using names arguments, the order of the arguments does not matter. We can also use what are called positional arguments, where `R` determines the meaning of the arguments based on their position.

```
plot(a, b)
```

This code does the same as the previous code. The expected position of arguments can be found in the help file for the function you are working with or by asking `R` to list the arguments using the `args` function.

```
args(plot)

## function (x, y, ...)
## NULL
```

It usually makes sense to use the positional arguments for only the first few arguments in a function. After that, named arguments are easier to keep track of. Many functions also have default argument values that will be used if values are not specified in the function call. These default argument values can be seen by using the `args` function and can also be found in the help files. For example, for the function `rnorm`, the arguments `mean` and `sd` have default values.

```
args(rnorm)

## function (n, mean = 0, sd = 1)
## NULL
```

Any time you want to call up a function, you must include parentheses after it, even if you are not specifying any arguments. If you do not include parentheses, R will return the function code (which at times might actually be useful).

Note that it is not necessary to use explicit numerical values as function arguments—symbolic variable names which represent appropriate data structure can be used. it is also possible to use functions as arguments within functions. R will evaluate such expressions from the inside outward. While this may seem trivial, this quality makes R very flexible. There is no explicit limit to the degree of nesting that can be used. You could use:

```
plot(rnorm(10, sqrt(mean(c(1:5, 7, 1, 8, sum(8.4, 1.2, 7))))), 1:10)
```

The above code includes 5 levels of nesting (the sum of 8.4,1.2 and 7 is combined with the other values to form a vector, for which the mean is calculated, then the square root of this value is taken and used as the standard deviation in a call to `rnorm`, and the output of this call is plotted). Of course, it is often easier to assign intermediate steps to symbolic variables. R evaluates nested expressions based on the values that functions return or the data represented by symbolic variables. For example, if a function expects character data for a particular argument, then you can use a call to the function `paste` in place of explicit character data.

Many functions (including `sum`, `plot`, and `rnorm`) come with the R "base packages", i.e., they are loaded and ready to go as soon as you open R. These packages contain the most common functions. While the base packages include many useful functions, for specialized procedures, you should check out the content that is available in the add-on packages. The CRAN website currently lists more than 4500 contributed packages that contain functions and data that users have contributed. You can find a list of the available packages at the CRAN website http://cran.r-project.org/. During the course of this book and described in more detail later on, we will be looking and using a number of specialized packages for application of DSM. Another repository of R packages is the R-Forge website https://r-forge.r-project.org/. R-Forge offers a central platform for the development of R packages, R-related software and further projects. Packages in R-Forge are not necessarily always on the CRAN website. However, many packages on the CRAN website are developed in R-Forge as ongoing projects. Sometimes to get the latest changes made upon a package, it pays to visit R-Forge first, as the uploading of the revised functions to CRAN is not instantaneous.

To utilize the functions in contributed R packages, you first need to install and then load the package. Packages can be installed via the packages menu in the right bottom panel of RStudio (select the "packages" menu, then "install packages"). Installation could be retrieved from the nearest mirror site (CRAN server location)—you will need to have first selected this by going to the *tools*, then *options*, then *packages* menu where you can then select the nearest mirror site from a suite of possibles. Alternatively, you may just install a package from a local zip file. This is fine, but often when using a package, there are other peripheral packages (or dependencies) that also need to be loaded (and installed). If you install the package from CRAN or a mirror site, the dependency packages are also installed. This is not the case when you are installing packages from zip files—you will also have to manually install all the dependencies too.

Or just use the command:

```
install.packages("package name")
```

where "package name" should be replaced with the actual name of the package you want to install, for example:

```
install.packages("Cubist")
```

This command will install the package of functions for running the Cubist rule-based machine learning models for regression.

Installation is a one-time process, but packages must be loaded each time you want to use them. This is very simple, e.g., to load the package `Cubist`, use the following command.

```
library(Cubist)
```

Similarly, if you want to install an R package from R-Forge (another popular hosting repository for R packages) you would use the following command:

```
install.packages("package name",repos="http://R-Forge.R-project.org")
```

Other popular repositories for R packages include Github and BitBucket. These repositories as well as R-Forge are version control systems that provide a central place for people to collaborate on everything from small to very large projects with speed and efficiency. The companion R package to this book, *ithir* is hosted on Bitbucket for example. *ithir* contains most of the data, and some important functions that are covered in this book so that users can replicate all of the analyses contained within. *ithir* can be downloaded and installed on your computer using the following commands:

```
library(devtools)
install_bitbucket("brendo1001/ithir/pkg")
library(ithir)
```

The above commands assumes your have already installed the devtools package. Any package that you want to use that is not included as one of the "base" packages, needs to be loaded every time you start R. Alternatively, you can add code to the file *Rprofile.site* that will be executed every time you start R.

You can find information on specific packages through CRAN, by browsing to http://cran.r-project.org/ and selecting the *packages* link. Each package has a separate web page, which will include links to source code, and a pdf manual. In RStudio, you can select the *packages* tab on the lower right panel. You will then see all the package that are currently installed in your R environment. By clicking onto any package, information on the various functions contained in the package, plus documentation and manuals for their usage. It becomes quite clear that within this RStudio environment, there is at your fingertips, a wealth of information for which to consult whenever you get stuck. When working with a new package, it is a good idea to read the manual.

To "unload" functions, use the detach function:

```
detach("package:Cubist")
```

For tasks that you repeat, but which have no associated function in R, or if you do not like the functions that are available, you can write your own functions. This will be covered a little a bit later on. Perhaps one day you may be able to compile all your functions that you have created into a R package for everyone else to use.

2.2.10 Getting Help

It is usually easy to find the answer about specific functions or about R in general. There are several good introductory books on R. For example, "R for Dummies", which has had many positive reviews http://www.amazon.com/R-Dummies-Joris-Meys/dp/1119962846.You can also find free detailed manuals on the CRAN website. Also, it helps to keep a copy of the "R Reference Card", which demonstrates the use of many common functions and operators in 4 pages http://cran.r-project.org/doc/contrib/Short-refcard.pdf. Often a Google search https://www.google.com.au/ of your problem can be a very helpful and fruitful exercise. To limit the results to R related pages, adding "cran" generally works well. R even has an internet search engine of sorts called *rseek*, which can be found at http://rseek.org/—it is really just like the Google search engine, but just for R stuff!

Each function in R has a help file associated with it that explains the syntax and usually includes an example. Help files are concisely written. You can bring up a help file by typing ? and then the function name.

```
>?cubist
```

This will bring up the help file for the cubist function in the help panel of RStudio. But, what if you are not sure what function you need for a particular task? How can you know what help file to open? In addition to the sources given below, you should try `help.search("keyword")` or `??keyword`, both of which search the R help files for whatever keyword you put in.

```
>??polygon
```

This will bring up a search results page in the help panel of RStudio of all the various help files that have something to do with `polygon`. In this case, i am only interested in a function that assesses whether a point is situated with a polygon. So looking down the list, one can see (provided the "SDMTools" package is installed) a function called `pnt.in.poly`. Clicking on this function, or submitting `?pnt.in.poly` to R will bring up the necessary help file.

There is an R help mailing list http://www.r-project.org/mail.html, which can be very helpful. Before posting a question, be sure to search the mailing list archives, and check the posting guide http://www.r-project.org/posting-guide.html.

One of the best sources of help on R functions is the mailing list archives (urlhttp://cran.r-project.org/, then select "search", then "searchable mail archives"). Here you can find suggestions for functions for particular problems, help on using specific functions, and all kinds of other information. A quick way to search the mailing list archives is by entering:

```
RSiteSearch("A Keyword")
```

For one more trick, to search for objects (including functions) that include a particular string, you can use the `apropos` function:

```
apropos("mean")

##  [1] ".colMeans"     ".rowMeans"     "colMeans"      "kmeans"
##  [5] "mean"          "mean.Date"     "mean.default"  "mean.difftime"
##  [9] "mean.POSIXct"  "mean.POSIXlt"  "rowMeans"      "weighted.mean"
```

2.2.11 Exercises

1. You can use for magic tricks: Pick any number. Double it, and then add 12 to the result. Divide by 2, and then subtract your original number. Did you end up with 6.0?
2. If you want to work with a set of 10 numbers in R, something like this:
   11 8.3 9.8 9.6 11.0 12.0 8.5 9.9 10.0 11.0
   - What type of data structure should you use to store these in R?
   - What if you want to work with a data set that contains site names, site locations, soil categorical information, soil property information, and some terrain variables—what type of data structure should you use to store these in R?
3. Install and load a package—take a look at the list of available packages, and pick one. To make sure you have loaded it correctly, try to run an example from

the package reference manual. Identify the arguments required for calling up the function. Detach the package when you are done.
4. Assign your full name to a variable called `my.name`. Print the value of `my.name`. Try to subtract 10 from `my.name`. Finally determine the type of data stored in `my.name` and 10 using the `class` function. If you are unsure of what `class` does, check out the help file.
5. You are interested in seeing what functions `R` has for fitting variograms (or some other topic of your choosing). Can you figure out how to search for relevant functions? Are you able to identify a function or two that may do what you want.

2.3 Vectors, Matrices, and Arrays

2.3.1 *Creating and Working with Vectors*

There are several ways to create a vector in `R`. Where the elements are spaced by exactly 1, just separate the values of the first and last elements with a colon.

```
1:5

## [1] 1 2 3 4 5
```

The function `seq` (for sequence) is more flexible. Its typical arguments are `from`, `to`, and `by` (or, in place of `by`, you can specify `length.out`).

```
seq(-10, 10, 2)

##  [1] -10  -8  -6  -4  -2   0   2   4   6   8  10
```

Note that the `by` argument does not need to be an integer. When all the elements in a vector are identical, use the `rep` function (for repeat).

```
rep(4, 5)

## [1] 4 4 4 4 4
```

For other cases, use `c` (for concatenate or combine).

```
c(2, 1, 5, 100, 2)

## [1]   2   1   5 100   2
```

Note that you can name the elements within a vector.

```
c(a = 2, b = 1, c = 5, d = 100, e = 2)

##   a   b   c   d   e
##   2   1   5 100   2
```

Any of these expressions could be assigned to a symbolic variable, using an assignment operator.

```
v1 <- c(2, 1, 5, 100, 2)
v1

## [1]   2   1   5 100   2
```

Variable names can be any combination of letters, numbers, and the symbols . and _, but they can not start with a number or with _. Google has a R style guide http://google-styleguide.googlecode.com/svn/trunk/google-r-style.html which describes good and poor examples of variable name attribution, but generally it is a personal preference on how you name your variables.

```
probably.not_a.good_example.for.a.name.100 <- seq(1, 2, 0.1)
probably.not_a.good_example.for.a.name.100

##  [1] 1.0 1.1 1.2 1.3 1.4 1.5 1.6 1.7 1.8 1.9 2.0
```

The c function is very useful for setting up arguments for other functions, as will be shown later. As with all R functions, both variable names and function names can be substituted into function calls in place of numeric values.

```
x <- rep(1:3)
y <- 4:10
z <- c(x, y)
z

##  [1]  1  2  3  4  5  6  7  8  9 10
```

Although R prints the contents of individual vectors with a horizontal orientation, R does not have "column vectors" and "row vectors", and vectors do not have a fixed orientation. This makes use of vectors in R very flexible.

Vectors do not need to contain numbers, but can contain data with any of the modes mentioned earlier (numeric, logical, character, and complex), as long as all the data in a vector are of the same mode.

Logical vectors are very useful in R for sub-setting data i.e., for isolating some part of an object that meets certain criteria. For relational commands, the shorter vector is repeated as many as necessary to carry out the requested comparison for each element in the longer vector.

```
x <- 1:10
x > 5

##  [1] FALSE FALSE FALSE FALSE FALSE  TRUE  TRUE  TRUE TRUE TRUE
```

Also, note that when logical vectors are used in arithmetic, they are changed (coerced in R terms) into a vector of binary elements: 1 or 0. Continuing with the above example:

```
a <- x > 5
a
```

```
##  [1] FALSE FALSE FALSE FALSE FALSE  TRUE  TRUE  TRUE TRUE TRUE
```

```
a * 1.4
```

```
##  [1] 0.0 0.0 0.0 0.0 0.0 1.4 1.4 1.4 1.4 1.4
```

One function that is commonly used on character data is `paste`. It concatenates character data (and can also work with numerical and logical elements—these become character data).

```
paste("A", "B", "C", "D", TRUE, 42)
```

```
## [1] "A B C D TRUE 42"
```

Note that the `paste` function is very different from `c`. The `paste` function concatenates its arguments into a single character value, while the `c` function combines its arguments into a vector, where each argument becomes a single element. The `paste` function becomes handy when you want to combine the character data that are stored in several symbolic variables.

```
month <- "April"
day <- 29
year <- 1770
paste("Captain Cook, on the ", day, "th day of ", month, ", "
, year, ", sailed into Botany Bay", sep = "")
```

```
## [1] "Captain Cook, on the 29th day of April, 1770,
##         sailed into Botany Bay"
```

This is especially useful with loops, when a variable with a changing value is combined with other data. Loops will be discussed in a later section.

```
group <- 1:10
id <- LETTERS[1:10]
for (i in 1:10) {
    print(paste("group =", group[i], "id =", id[i]))
}
```

```
## [1] "group = 1 id = A"
## [1] "group = 2 id = B"
## [1] "group = 3 id = C"
## [1] "group = 4 id = D"
## [1] "group = 5 id = E"
## [1] "group = 6 id = F"
## [1] "group = 7 id = G"
## [1] "group = 8 id = H"
## [1] "group = 9 id = I"
## [1] "group = 10 id = J"
```

`LETTERS` is a constant that is built into `R`—it is a vector of uppercase letters A through Z (different from `letters`).

2.3.2 Vector Arithmetic, Some Common Functions, and Vectorised Operations

In R, vectors can be used directly in arithmetic operations. Operations are applied on an element-by-element basis. This can be referred to as "vectorised" arithmetic, and along with vectorised functions (described below), it is a quality that makes R a very efficient programming language.

```
x <- 6:10
x

## [1]  6  7  8  9 10

x + 2

## [1]  8  9 10 11 12
```

For an operation carried out on two vectors, the mathematical operation is applied on an element-by-element basis.

```
y <- c(4, 3, 7, 1, 1)
y

## [1] 4 3 7 1 1

z <- x + y
z

## [1] 10 10 15 10 11
```

When two vectors having different numbers of elements used in an expression together, R will repeat the smaller vector. For example, with vector of length one, i.e., a single number:

```
x <- 1:10
m <- 0.8
b <- 2
y <- m * x + b
y

##  [1]  2.8  3.6  4.4  5.2  6.0  6.8  7.6  8.4  9.2 10.0
```

If the number of rows in the smaller vector is not a multiple of the larger vector (often indicative of an error) R will return a warning.

```
x <- 1:10
m <- 0.8
b <- c(2, 1, 1)
y <- m * x + b
```

```
## Warning in m * x + b: longer object length is not a multiple
of shorter object length
```

```
y
```

```
##  [1]  2.8  2.6  3.4  5.2  5.0  5.8  7.6  7.4  8.2 10.0
```

Some arithmetic operators that are available in R include:

`+`	addition
`-`	subtraction
`*`	multiplication
`/`	division
`^`	exponentiation
`%/%`	integer division
`%%`	modulo (remainder)
`log (a)`	natural log of a
`log10 (a)`	base 10 log of a
`exp (a)`	e^a
`sine (a)`	sine of a
`cos (a)`	cosine of a
`tan (a)`	tangent of a
`sqrt (a)`	square root of a

Some simple functions that are useful for vector math include:

`min`	minimum value of a set of numbers
`max`	maximum of a set of numbers
`pmin`	parallel minima (compares multiple vectors "row-by-row")
`pmax`	parallel maxima
`sum`	sum of all elements
`length`	length of a vector (or number of columns in a data frame)
`nrow`	number of rows in a vector of data frame
`ncol`	number of columns
`mean`	arithmetic mean
`sd`	standard deviation
`rnorm`	generates a vector of normally-distributed random numbers
`signif, ceiling, floor`	rounding

Many, many other functions are available.

R also has a few built in constants, including `pi`.

```
pi
```

```
## [1] 3.141593
```

Parentheses can be used to control the order of operations, as in any other programming language.

```
7 - 2 * 4

## [1] -1
```

is different from:

```
(7 - 2) * 4

## [1] 20
```

and

```
10^1:5

## [1] 10  9  8  7  6  5
```

is different from:

```
10^(1:5)

## [1] 1e+01 1e+02 1e+03 1e+04 1e+05
```

Many functions in R are capable of accepting vectors (or even data frames, arrays, and lists) as input for single arguments, and returning an object with the same structure. These vectorised functions make vector manipulations very efficient. Examples of such functions include `log`, `sin`, and `sqrt`, For example:

```
x <- 1:10
sqrt(x)

##  [1] 1.000000 1.414214 1.732051 2.000000 2.236068 2.449490 2.645751
##  [8] 2.828427 3.000000 3.162278
```

or

```
sqrt(1:10)

##  [1] 1.000000 1.414214 1.732051 2.000000 2.236068 2.449490 2.645751
##  [8] 2.828427 3.000000 3.162278
```

The previous expressions are also equivalent to:

```
sqrt(c(1, 2, 3, 4, 5, 6, 7, 8, 9, 10))

##  [1] 1.000000 1.414214 1.732051 2.000000 2.236068 2.449490 2.645751
##  [8] 2.828427 3.000000 3.162278
```

But they are not the same as the following, where all the numbers are interpreted as individual values for multiple arguments.

```
sqrt(1, 2, 3, 4, 5, 6, 7, 8, 9, 10)

## Error in sqrt(1, 2, 3, 4, 5, 6, 7, 8, 9, 10): 10 arguments
passed to 'sqrt' which requires 1
```

There are also some functions designed for making vectorised operations on lists, matrices, and arrays: these include `apply` and `lapply`.

2.3.3 *Matrices and Arrays*

Arrays are multi-dimensional collections of elements and matrices are simply two-dimensional arrays. R has several operators and functions for carrying out operations on arrays, and matrices in particular (e.g., matrix multiplication).

To generate a matrix, the `matrix` function can be used. For example:

```
X <- matrix(1:15, nrow = 5, ncol = 3)
X

##      [,1] [,2] [,3]
## [1,]    1    6   11
## [2,]    2    7   12
## [3,]    3    8   13
## [4,]    4    9   14
## [5,]    5   10   15
```

Note that the filling order is by column by default (i.e., each column is filled before moving onto the next one). The "unpacking" order is the same:

```
as.vector(X)

##  [1]  1  2  3  4  5  6  7  8  9 10 11 12 13 14 15
```

If, for any reason, you want to change the filling order, you can use the `by row` argument:

```
X <- matrix(1:15, nrow = 5, ncol = 3, byrow = T)
X

##      [,1] [,2] [,3]
## [1,]    1    2    3
## [2,]    4    5    6
## [3,]    7    8    9
## [4,]   10   11   12
## [5,]   13   14   15
```

A similar function is available for higher-order arrays, called `array`. Here is an example with a three-dimensional array:

```
Y <- array(1:30, dim = c(5, 3, 2))
Y

## , , 1
##
##      [,1] [,2] [,3]
## [1,]    1    6   11
## [2,]    2    7   12
## [3,]    3    8   13
## [4,]    4    9   14
## [5,]    5   10   15
##
## , , 2
##
##      [,1] [,2] [,3]
## [1,]   16   21   26
## [2,]   17   22   27
## [3,]   18   23   28
## [4,]   19   24   29
## [5,]   20   25   30
```

Arithmetic with matrices and arrays that have the same dimensions is straightforward, and is done on an element-by-element basis. This true for all the arithmetic operators listed in earlier sections.

```
Z <- matrix(1, nrow = 5, ncol = 3)
Z

##      [,1] [,2] [,3]
## [1,]    1    1    1
## [2,]    1    1    1
## [3,]    1    1    1
## [4,]    1    1    1
## [5,]    1    1    1

X + Z

##      [,1] [,2] [,3]
## [1,]    2    3    4
## [2,]    5    6    7
## [3,]    8    9   10
## [4,]   11   12   13
## [5,]   14   15   16
```

This does not work when dimensions do not match:

```
Z <- matrix(1, nrow = 3, ncol = 3)
X + Z

## Error in X + Z: non-conformable arrays
```

For mixed vector/array arithmetic, vectors are recycled if needed.

```
Z
```

```
##      [,1] [,2] [,3]
## [1,]    1    1    1
## [2,]    1    1    1
## [3,]    1    1    1
```

```
x <- 1:9
Z + x
```

```
##      [,1] [,2] [,3]
## [1,]    2    5    8
## [2,]    3    6    9
## [3,]    4    7   10
```

```
y <- 1:3
Z     + y
```

```
##      [,1] [,2] [,3]
## [1,]    2    2    2
## [2,]    3    3    3
## [3,]    4    4    4
```

R also has operators for matrix algebra. The operator `%*%` carries out matrix multiplication, and the function `solve` can invert matrices.

```
X <- matrix(c(1, 2.5, 6, 7.5, 4.9, 5.6, 9.9, 7.8, 9.3), nrow = 3)
X
```

```
##      [,1] [,2] [,3]
## [1,]  1.0  7.5  9.9
## [2,]  2.5  4.9  7.8
## [3,]  6.0  5.6  9.3
```

```
solve(X)
```

```
##             [,1]       [,2]       [,3]
## [1,]  0.07253886 -0.5492228  0.3834197
## [2,]  0.90385723 -1.9228555  0.6505469
## [3,] -0.59105738  1.5121858 -0.5315678
```

2.3.4 Exercises

1. Generate a vector of numbers that contains the sequence 1, 2, 3, … 10 (try to use the least amount of code possible to do this). Assign this vector to the variable x, and then carry out the following vector arithmetic.

   (a) $log_{10}x$
   (b) lnx
   (c) $\frac{\sqrt{x}}{2-x}$

2. Use an appropriate function to generate a vector of 100 numbers that go from 0 to 2π, with a constant interval. Assuming this fist vector is called x, create a new vector that contains $sine(2x - 0.5\pi)$. Determine the minimum and maximum of $sine(2x - 0.5\pi)$. Does this match with what you expect?
3. Create 5 vectors, each containing 10 random numbers. Give each vector a different name. Create a new vector where the 1st element contains the sum of the 1st elements in your original 5 vectors, the 2nd element contains the sum of the 2nd elements, etc. Determine the mean of this new vector.
4. Create the following matrix using the least amount of code:

```
##      [,1] [,2] [,3] [,4] [,5]
## [1,]    1    2    3    4    5
## [2,]    6    7    8    9   10
## [3,]   11   12   13   14   15
## [4,]   16   17   18   19   20
## [5,]   21   22   23   24   25
```

5. If you are bored, try this. Given the following set of linear equations:

$$27.2x + 32y - 10.8z = 401.2$$
$$x - 1.48y = 0$$
$$409.1x + 13.5z = 2.83$$

Solve for x, y, and z.

2.4 Data Frames, Data Import, and Data Export

As described above, a data frame is a type of data structure in R with rows and columns, where different columns contain data with different modes. A data frame is probably the most common data structure that you will use for storing soil information data sets. Recall from before the data frame that we created.

```
dat <- data.frame(profile_id = c("Chromosol", "Vertosol", "Sodosol"),
FID = c("a1", "a10", "a11"), easting = c(337859, 344059, 347034),
northing = c(6372415, 6376715, 6372740), visted = c(TRUE, FALSE, TRUE))
dat
```

```
##   profile_id FID easting northing visted
## 1  Chromosol  a1  337859  6372415   TRUE
## 2   Vertosol a10  344059  6376715  FALSE
## 3    Sodosol a11  347034  6372740   TRUE
```

We can quickly assess the different modes of a data frame (or any other object for that matter) by using the `str`(structure) function.

```
str(dat)
```

```
## 'data.frame': 3 obs. of  5 variables:
```

```
##  $ profile_id: Factor w/ 3 levels "Chromosol","Sodosol",..:
##  $ FID       : Factor w/ 3 levels "a1","a10","a11": 1 2 3
##  $ easting   : num  337859 344059 347034
##  $ northing  : num  6372415 6376715 6372740
##  $ visted    : logi  TRUE FALSE TRUE
```

The `str` is probably one of the most used `R` functions. It is great for exploring the format and contents of any object created or imported.

2.4.1 Reading Data from Files

The easiest way to create a data frame is to read in data from a file—this is done using the function `read.table`, which works with ASCII text files. Data can be read in from other files as well, using different functions, but `read.table` is the most commonly used approach. `R` is very flexible in how it reads in data from text files.

Note that the column labels in the header have to be compatible with `R`'s variable naming convention, or else `R` will make some changes as they are read in (or will not read in the data correctly). So lets import some real soil data. These soil data are some chemical and physical properties from a collection of soil profiles sampled at various locations in New South Wales, Australia

```
soil.data <- read.table("USYD_soil1.txt", header = TRUE, sep = ",")

## Warning in file(file, "rt"): cannot open file
'USYD_soil1.txt': No such file or directory

## Error in file(file, "rt"): cannot open the connection

str(soil.data)

## Error in str(soil.data): object 'soil.data' not found

head(soil.data)

## Error in head(soil.data): object 'soil.data' not found
```

However, you may find that an error occurs, saying something like that the file does not exist. This is true as it has not been provided to you. Rather, to use this data you will need to load up the previously installed `ithir` package.

```
library(ithir)
data(USYD_soil1)
soil.data <- USYD_soil1
str(soil.data)

## 'data.frame': 166 obs. of  16 variables:
```

```
## $ PROFILE      : int  1 1 1 1 1 1 2 2 2 2 ...
## $ Landclass    : Factor w/ 4 levels "Cropping","Forest",..: 4 4 4 ...
## $ Upper.Depth : num  0 0.02 0.05 0.1 0.2 0.7 0 0.02 0.05 0.1 ...
## $ Lower.Depth : num  0.02 0.05 0.1 0.2 0.3 0.8 0.02 0.05 0.1 0.2 ...
## $ clay         : int  8 8 8 8 NA 57 9 9 9 NA ...
## $ silt         : int  9 9 10 10 10 8 10 10 10 10 ...
## $ sand         : int  83 83 82 83 79 36 81 80 80 81 ...
## $ pH_CaCl2     : num  6.35 6.34 4.76 4.51 4.64 6.49 5.91 5.94 5.63 4.22 ...
## $ Total_Carbon: num  1.07 0.98 0.73 0.39 0.23 0.35 1.14 1.14 1.01 0.48 ...
## $ EC           : num  0.168 0.137 0.072 0.034 NA 0.059 0.123 0.101...
## $ ESP          : num  0.3 0.5 0.9 0.2 0.9 0.3 0.3 0.6 NA NA ...
## $ ExchNa       : num  0.01 0.02 0.02 0 0.02 0.04 0.01 0.02 NA NA ...
## $ ExchK        : num  0.71 0.47 0.52 0.38 0.43 0.46 0.7 0.56 NA NA ...
## $ ExchCa       : num  3.17 3.5 1.34 1.03 1.5 9.13 2.92 3.2 NA NA ...
## $ ExchMg       : num  0.59 0.6 0.22 0.22 0.5 5.02 0.51 0.5 NA NA ...
## $ CEC          : num  5.29 3.7 2.86 2.92 2.6 ...
```

```
head(soil.data)
```

```
##   PROFILE      Landclass Upper.Depth Lower.Depth clay silt sand pH_CaCl2
## 1        1 native pasture        0.00        0.02    8    9   83     6.35
## 2        1 native pasture        0.02        0.05    8    9   83     6.34
## 3        1 native pasture        0.05        0.10    8   10   82     4.76
## 4        1 native pasture        0.10        0.20    8   10   83     4.51
## 5        1 native pasture        0.20        0.30   NA   10   79     4.64
## 6        1 native pasture        0.70        0.80   57    8   36     6.49
##   Total_Carbon    EC ESP ExchNa ExchK ExchCa ExchMg   CEC
## 1         1.07 0.168 0.3   0.01  0.71   3.17   0.59  5.29
## 2         0.98 0.137 0.5   0.02  0.47   3.50   0.60  3.70
## 3         0.73 0.072 0.9   0.02  0.52   1.34   0.22  2.86
## 4         0.39 0.034 0.2   0.00  0.38   1.03   0.22  2.92
## 5         0.23    NA 0.9   0.02  0.43   1.50   0.50  2.60
## 6         0.35 0.059 0.3   0.04  0.46   9.13   5.02 14.96
```

When we import a file into R, it is good habit to look at its structure (`str`) to ensure the data is as it should be. As can be seen, this data set (frame) has 166 observations, and 15 columns. The `head` function is also a useful exploratory function, which simply allows us to print out the data frame, but only the first 6 rows of it (good for checking data frame integrity). Note that you must specify `header=TRUE`, or else R will interpret the row of labels as data. If the file you are loading is not in the directory that R is working in (the working directory, which can be checked with `getwd()` and changed with `setwd(file = "filename")`). When setting the working directory (`setwd()`), you can include the file path, but note that the path should have forward, not backward slashes (or double backward slashes, if you prefer).

The column separator function `sep` (an argument of `read.table`) lets you tell R, where the column breaks or delimiters occur. In the `soil.data` object, we specify that the data is comma separated. If you do not specify a field separator, R assumes that any spaces or tabs separate the data in your text file. However, any character data that contain spaces must be surrounded by quotes (otherwise, R interprets the data on either side of the white spaces as different elements).

Besides comma separators, tab separators are also common `sep="ˆ'`, so is `sep=".″` (full stop separator). Two consecutive separators will be interpreted as a missing value. Conversely, with the default options, you need to explicitly identify missing values in your data file with `NA` (or any other character, as long as you tell `R` what it is with the `na.strings` argument).

For some field separators, there are alternate functions that can be used with the default arguments, e.g., `read.csv`, which is identical to `read.table`, except default arguments differ. Also, `R` does not care what the name of your file is, or what its extension is, as long as it is an ASCII text file. A few other handy bits of information for using `read.table` follow. You can include comments at the end of rows in your data file—just precede them with a `#`. Also `R` will recognize `NaN`, `Inf`, and `-Inf` in input files.

Probably the easiest approach to handling missing values is to indicate their presence with `NA` in the text file. `R` will automatically recognize these as missing values. Alternatively, just leave the missing values as they are (blank spaces), and let `R` do the rest.

```
which(is.na(soil.data$CEC))

## [1]   9  10  45  63 115

soil.data[8:11, ]

##    PROFILE          Landclass Upper.Depth Lower.Depth clay silt sand
## 8        2 improved pasture        0.02        0.05    9   10   80
## 9        2 improved pasture        0.05        0.10    9   10   80
## 10       2 improved pasture        0.10        0.20   NA   10   81
## 11       2 improved pasture        0.38        0.50   36    8   56
##    pH_CaCl2 Total_Carbon    EC ESP ExchNa ExchK ExchCa ExchMg CEC
## 8      5.94         1.14 0.101 0.6   0.02  0.56    3.2    0.5 4.0
## 9      5.63         1.01 0.026  NA     NA    NA     NA     NA  NA
## 10     4.22         0.48 0.042  NA     NA    NA     NA     NA  NA
## 11     6.48         0.18 0.053 0.4   0.04  0.49    6.0    2.3 8.6
```

In most cases, it makes sense to put your data into a text file for reading into `R`. This can be done in various ways. Data download from the internet are often in text files to begin with. Data can be entered directly into a text file using a text editor. For data that are in spreadsheet program such as Excel or JMP, there are facilities available for saving these tabular frames to text files for reading into `R`.

This all may seem confusing, but it is really not that bad. Your best bet is to play around with the different options, find one that you like, and stick with it. Lastly, data frames can also be edited interactively in `R` using the `edit` function. This is really only useful for small data sets, and the function is not supported by RStudio (you could try with using Tinn-R instead if you want to explore using this function)

```
soil.data <- edit(soil.data)
```

2.4.2 *Creating Data Frames Manually*

Data frames can be made manually using the `data.frame` function:

```
soil <- c("Chromosol", "Vertosol", "Organosol", "Anthroposol")
carbon <- c(2.1, 2.9, 5.5, 0.2)
dat <- data.frame(soil.type = soil, soil.OC = carbon)
dat
```

```
##     soil.type soil.OC
## 1   Chromosol     2.1
## 2    Vertosol     2.9
## 3   Organosol     5.5
## 4 Anthroposol     0.2
```

While this approach is not an efficient way to enter data that could be read in directly, it can be very handy for some applications, such as the creation of customized summary tables. Note that column names are specified using an equal sign. It is also possible to specify (or change, or check) column names for an existing data frame using the function `names`.

```
names(dat) <- c("soil", "SOC")
dat
```

```
##          soil SOC
## 1   Chromosol 2.1
## 2    Vertosol 2.9
## 3   Organosol 5.5
## 4 Anthroposol 0.2
```

Row names can be specified in the `data.frame` function with the `row.names` argument.

```
dat <- data.frame(soil.type = soil, soil.OC = carbon,
row.names = c("Ch", "Ve", "Or", "An"))
dat
```

```
##      soil.type soil.OC
## Ch   Chromosol     2.1
## Ve    Vertosol     2.9
## Or   Organosol     5.5
## An Anthroposol     0.2
```

Specifying row names can be useful if you want to index data, which will be covered later. Row names can also be specified for an existing data frame with the `rownames` function (not to be confused with the `row.names` argument).

2.4.3 Working with Data Frames

So what do you do with data in R once it is in a data frame? Commonly, the data in a data frame will be used in some type of analysis or plotting procedure. It is usually necessary to be able to select and identify specific columns (i.e., vectors) within data frames. There are two ways to specify a given column of data within a data frame. The first is to use the $ notation. To see what the column names are, we can use the names function. Using our soil.data set:

```
names(soil.data)

##  [1] "PROFILE"       "Landclass"     "Upper.Depth"   "Lower.Depth"
##  [5] "clay"          "silt"          "sand"          "pH_CaCl2"
##  [9] "Total_Carbon"  "EC"            "ESP"           "ExchNa"
## [13] "ExchK"         "ExchCa"        "ExchMg"        "CEC"
```

The $ just uses a $ between the data frame and column name to specify a particular column. Say we want to look at the ESP column, which is the acronym for exchangeable sodium percentage.

```
soil.data$ESP

##   [1]  0.3  0.5  0.9  0.2  0.9  0.3  0.3  0.6   NA   NA  0.4  0.9  0.2  0.1
##  [15]   NA  0.4  0.5  0.7  0.2  0.1   NA  0.2  0.3   NA  0.8  0.6  0.8  0.9
##  [29]  0.9  1.1  0.5  0.6  1.1  0.2  0.6  1.1  0.4  0.3  0.5  1.0  2.2  0.1
##  [43]  0.1  0.1   NA  0.1  0.1  0.4   NA  0.2  0.1  0.3  0.4  0.1  0.4  0.1
##  [57]  0.1   NA  0.1  0.3   NA  0.1   NA  0.2  1.8  2.6  0.2 13.0  0.0  0.1
##  [71]  0.3  0.1  0.1  0.3  0.0  0.3  0.6  0.9  0.4   NA   NA  2.4  0.2  0.3
##  [85]  0.2  0.0  0.1  0.4   NA  0.3  0.3  0.2  0.3  0.6  0.3  0.2  0.7  0.3
##  [99]  0.4  1.0  7.9  6.1  5.7  5.2  4.7  2.9  5.8  7.2  9.6   NA 17.4   NA
## [113] 11.1  6.4   NA  4.0 12.1 21.2  2.2  1.9   NA  4.0 13.2  0.9  0.8  0.5
## [127]  0.2  0.4   NA  1.2  0.6  0.2  1.0  0.4  0.6  0.1  0.4   NA  0.7  0.5
## [141]  0.7  0.9  4.8  3.8  4.9  6.2 10.4 16.4  2.7   NA  1.2  0.5  1.9  2.0
## [155]  2.1  1.9  1.8  3.5  7.7  2.7  1.8  0.8  0.5  0.5  0.3  0.9
```

Although it is handy to think of data frame columns to have a vertical orientation, this orientation is not present when they are printed individually—instead, elements are printed from left to right, and then top to bottom. The expression soil.data$ESP could be used just as you would for any other vector. For example:

```
mean(soil.data$ESP)

## [1] NA
```

R can not calculate the mean because of the NA values in the vector. Lets remove them first using the na.omit function.

```
mean(na.omit(soil.data$ESP))

## [1] 1.99863
```

The second option for working with individual columns within a data frame is to use the commands `attach` and `detach`. Both of these functions take a data frame as an argument. `attach`ing a data frame puts all the columns within the that data frame into `R`'s search path, and they can be called by using their names alone without the `$` notation.

```
attach(soil.data)
ESP
```

```
##   [1]  0.3  0.5  0.9  0.2  0.9  0.3  0.3  0.6   NA   NA  0.4  0.9  0.2  0.1
##  [15]   NA  0.4  0.5  0.7  0.2  0.1   NA  0.2  0.3   NA  0.8  0.6  0.8  0.9
##  [29]  0.9  1.1  0.5  0.6  1.1  0.2  0.6  1.1  0.4  0.3  0.5  1.0  2.2  0.1
##  [43]  0.1  0.1   NA  0.1  0.1  0.4   NA  0.2  0.1  0.3  0.4  0.1  0.4  0.1
##  [57]  0.1   NA  0.1  0.3   NA  0.1   NA  0.2  1.8  2.6  0.2 13.0  0.0  0.1
##  [71]  0.3  0.1  0.1  0.3  0.0  0.3  0.6  0.9  0.4   NA   NA  2.4  0.2  0.3
##  [85]  0.2  0.0  0.1  0.4   NA  0.3  0.3  0.2  0.3  0.6  0.3  0.2  0.7  0.3
##  [99]  0.4  1.0  7.9  6.1  5.7  5.2  4.7  2.9  5.8  7.2  9.6   NA 17.4   NA
## [113] 11.1  6.4   NA  4.0 12.1 21.2  2.2  1.9   NA  4.0 13.2  0.9  0.8  0.5
## [127]  0.2  0.4   NA  1.2  0.6  0.2  1.0  0.4  0.6  0.1  0.4   NA  0.7  0.5
## [141]  0.7  0.9  4.8  3.8  4.9  6.2 10.4 16.4  2.7   NA  1.2  0.5  1.9  2.0
## [155]  2.1  1.9  1.8  3.5  7.7  2.7  1.8  0.8  0.5  0.5  0.3  0.9
```

Note that when you are done using the individual columns, it is good practice to `detach` your data frame. Once the data frame is `detach`ed, `R` will no longer know what you mean when you specify the name of the column alone:

```
detach(soil.data)
ESP
```

```
## Error in eval(expr, envir, enclos): object 'ESP' not found
```

If you modify a variable that is part of an attached data frame, the data within the data frame remain unchanged; you are actually working with a copy of the data frame.

Another option (for selecting particular columns) is to use the square braces `[]` to specify the column you want. Using the square braces to select the `ESP` column from our data set you would use:

```
soil.data[, 10]
```

Here you are specifying the column in the tenth position, which as you should check is the `ESP` column. To use the square braces the row position precedes to comma, and the column position proceeds to comma. By leaving a blank space in front of the comma, we are essentially instruction `R` to print out the whole column. You may be able to surmise that it is also possible to subset a selection of columns quite efficiently with this square brace method. We will use the square braces more a little later on.

The `$` notation can also be used to add columns to a data frame. For example, if we want to express our `Upper.Depth` and `Lower.Depth` columns in *cm* rather than *m* we could do the following.

```
soil.data$Upper <- soil.data$Upper.Depth * 100
soil.data$Lower <- soil.data$Lower.Depth * 100
head(soil.data)
```

```
##   PROFILE      Landclass Upper.Depth Lower.Depth clay silt sand pH_CaCl2
## 1       1 native pasture        0.00        0.02    8    9   83     6.35
## 2       1 native pasture        0.02        0.05    8    9   83     6.34
## 3       1 native pasture        0.05        0.10    8   10   82     4.76
## 4       1 native pasture        0.10        0.20    8   10   83     4.51
## 5       1 native pasture        0.20        0.30   NA   10   79     4.64
## 6       1 native pasture        0.70        0.80   57    8   36     6.49
##   Total_Carbon    EC ESP ExchNa ExchK ExchCa ExchMg   CEC Upper Lower
## 1         1.07 0.168 0.3   0.01  0.71   3.17   0.59  5.29     0     2
## 2         0.98 0.137 0.5   0.02  0.47   3.50   0.60  3.70     2     5
## 3         0.73 0.072 0.9   0.02  0.52   1.34   0.22  2.86     5    10
## 4         0.39 0.034 0.2   0.00  0.38   1.03   0.22  2.92    10    20
## 5         0.23    NA 0.9   0.02  0.43   1.50   0.50  2.60    20    30
## 6         0.35 0.059 0.3   0.04  0.46   9.13   5.02 14.96    70    80
```

Many data frames that contain real data will have some missing observations. R has several tools for working with these observations. For starters, the `na.omit` function can be used for removing NAs from a vector. Working again with the `ESP` column of our `soil.data` set:

```
soil.data$ESP
```

```
##   [1]  0.3  0.5  0.9  0.2  0.9  0.3  0.3  0.6   NA   NA  0.4  0.9  0.2  0.1
##  [15]   NA  0.4  0.5  0.7  0.2  0.1   NA  0.2  0.3   NA  0.8  0.6  0.8  0.9
##  [29]  0.9  1.1  0.5  0.6  1.1  0.2  0.6  1.1  0.4  0.3  0.5  1.0  2.2  0.1
##  [43]  0.1  0.1   NA  0.1  0.1  0.4   NA  0.2  0.1  0.3  0.4  0.1  0.4  0.1
##  [57]  0.1   NA  0.1  0.3   NA  0.1   NA  0.2  1.8  2.6  0.2 13.0  0.0  0.1
##  [71]  0.3  0.1  0.1  0.3  0.0  0.3  0.6  0.9  0.4   NA   NA  2.4  0.2  0.3
##  [85]  0.2  0.0  0.1  0.4   NA  0.3  0.3  0.2  0.3  0.6  0.3  0.2  0.7  0.3
##  [99]  0.4  1.0  7.9  6.1  5.7  5.2  4.7  2.9  5.8  7.2  9.6   NA 17.4   NA
## [113] 11.1  6.4   NA  4.0 12.1 21.2  2.2  1.9   NA  4.0 13.2  0.9  0.8  0.5
## [127]  0.2  0.4   NA  1.2  0.6  0.2  1.0  0.4  0.6  0.1  0.4   NA  0.7  0.5
## [141]  0.7  0.9  4.8  3.8  4.9  6.2 10.4 16.4  2.7   NA  1.2  0.5  1.9  2.0
## [155]  2.1  1.9  1.8  3.5  7.7  2.7  1.8  0.8  0.5  0.5  0.3  0.9
```

```
na.omit(soil.data$ESP)
```

```
##   [1]  0.3  0.5  0.9  0.2  0.9  0.3  0.3  0.6  0.4  0.9  0.2  0.1  0.4  0.5
##  [15]  0.7  0.2  0.1  0.2  0.3  0.8  0.6  0.8  0.9  0.9  1.1  0.5  0.6  1.1
##  [29]  0.2  0.6  1.1  0.4  0.3  0.5  1.0  2.2  0.1  0.1  0.1  0.1  0.1  0.4
##  [43]  0.2  0.1  0.3  0.4  0.1  0.4  0.1  0.1  0.1  0.3  0.1  0.2  1.8  2.6
##  [57]  0.2 13.0  0.0  0.1  0.3  0.1  0.1  0.3  0.0  0.3  0.6  0.9  0.4  2.4
##  [71]  0.2  0.3  0.2  0.0  0.1  0.4  0.3  0.3  0.2  0.3  0.6  0.3  0.2  0.7
##  [85]  0.3  0.4  1.0  7.9  6.1  5.7  5.2  4.7  2.9  5.8  7.2  9.6 17.4 11.1
##  [99]  6.4  4.0 12.1 21.2  2.2  1.9  4.0 13.2  0.9  0.8  0.5  0.2  0.4  1.2
## [113]  0.6  0.2  1.0  0.4  0.6  0.1  0.4  0.7  0.5  0.7  0.9  4.8  3.8  4.9
## [127]  6.2 10.4 16.4  2.7  1.2  0.5  1.9  2.0  2.1  1.9  1.8  3.5  7.7  2.7
## [141]  1.8  0.8  0.5  0.5  0.3  0.9
## attr(,"na.action")
##  [1]   9  10  15  21  24  45  49  58  61  63  80  81  89 110 112 115 121
## [18] 129 138 150
## attr(,"class")
## [1] "omit"
```

Although the result does contain more than just the non-NA values, only the non-NA values will be used in subsequent operations. Note that the result of na.omit contains more information than just the non-NA values. This function can also be applied to complete data frames. In this case, any row with an NA is removed (so be careful with its usage).

```
soil.data.cleaned <- na.omit(soil.data)
```

It is often necessary to identify NAs present in a data structure. The is.na function can be used for this—it can also be negated using the "!" character.

```
is.na(soil.data$ESP)

##   [1] FALSE FALSE FALSE FALSE FALSE FALSE FALSE FALSE  TRUE  TRUE FALSE
##  [12] FALSE FALSE FALSE  TRUE FALSE FALSE FALSE FALSE FALSE  TRUE FALSE
##  [23] FALSE  TRUE FALSE FALSE FALSE FALSE FALSE FALSE FALSE FALSE FALSE
##  [34] FALSE FALSE FALSE FALSE FALSE FALSE FALSE FALSE FALSE FALSE FALSE
##  [45]  TRUE FALSE FALSE FALSE  TRUE FALSE FALSE FALSE FALSE FALSE FALSE
##  [56] FALSE FALSE  TRUE FALSE FALSE  TRUE FALSE  TRUE FALSE FALSE FALSE
##  [67] FALSE FALSE FALSE FALSE FALSE FALSE FALSE FALSE FALSE FALSE FALSE
##  [78] FALSE FALSE  TRUE  TRUE FALSE FALSE FALSE FALSE FALSE FALSE FALSE
##  [89]  TRUE FALSE FALSE FALSE FALSE FALSE FALSE FALSE FALSE FALSE FALSE
## [100] FALSE FALSE FALSE FALSE FALSE FALSE FALSE FALSE FALSE FALSE  TRUE
## [111] FALSE  TRUE FALSE FALSE  TRUE FALSE FALSE FALSE FALSE FALSE  TRUE
## [122] FALSE FALSE FALSE FALSE FALSE FALSE FALSE  TRUE FALSE FALSE FALSE
## [133] FALSE FALSE FALSE FALSE FALSE  TRUE FALSE FALSE FALSE FALSE FALSE
## [144] FALSE FALSE FALSE FALSE FALSE FALSE  TRUE FALSE FALSE FALSE FALSE
## [155] FALSE FALSE FALSE FALSE FALSE FALSE FALSE FALSE FALSE FALSE FALSE
## [166] FALSE
```

2.4.4 *Writing Data to Files*

With R, it is easy to write data to files. The function write.table is usually the best function for this purpose. Given only a data frame and a file name, this function will write the data contained in the data frame to a text file. There are a number of arguments that can be controlled with this function as shown below (also look at the help file).

write.table(x, file = "", append = FALSE, quote = TRUE, sep = " ", eol = "\", na = "NA", dec = ".", row.names = TRUE, col.names = TRUE, qmethod = c("escape", "double"), fileEncoding = "")

The important ones (or most frequently used) are the column separator sep argument, and whether of not you want to keep the column and row names (col.names and row.names respectively). For example, if we want to write soil.data to text file ("file name.txt"), retaining the column names, not retaining row names, and having a tab delimited column separator, we would use:

```
write.table(soil.data, file = "file name.txt", col.names = TRUE,
row.names = FALSE, sep = "\t")
```

Setting the append argument to TRUE lets you add data to a file that already exists.

The write.table function can not be used with all data structures in R (like lists for example). However, it can be used for such things as vectors and matrices.

2.4.5 Exercises

1. Using the soil.data object, determine the minimum and maximum soil pH (PH_CaCl2) in the data frame. Next add a new column to the dataframe that contains the $log_1 0$ of soil carbon Total_Carbon.
2. Create a new data frame that contains the mean SOC, pH, and clay of the data set. Write out the summary to a new file using the default options. Finally, try changing the separator to a tab and write to a new file.
3. There are a number of NA values in the data set. We want to remove them. Could this be done in one step i.e., delete every row that contains an NA? Is this appropriate? How would you go about ensuring that no data is lost? Can you do this? or perhaps—do this!

2.5 Graphics: The Basics

2.5.1 *Introduction to the* Plot *Function*

It is easy to produce publication-quality graphics in R. There are many excellent R packages at your finger tips to do this; some of which include lattice and ggplot2 (see the help files and documentation for these). While in the course of this book we will revert to using these "high end" plotting packages, some fundamentals of plotting need to bedded down. Therefore in this section we will focus on the simplest plots—those which can be produced using the plot function, which is a base function that come with R. This function produces a plot as a side effect, but the type of plot produced depends on the type of data submitted. The basic plot arguments, as given in the help file for plot.default are:

```
plot(x, y = NULL, type = ''p'', xlim = NULL, ylim = NULL, log =
'''', main = NULL, sub = NULL, xlab = NULL, ylab = NULL, ann =
par(''ann''), axes = TRUE, frame.plot = axes, panel.first = NULL,
panel.last = NULL, asp = NA, ...)
```

To plot a single vector, all we need to do is supply that vector as the only argument to the function. This plot is shown in Fig. 2.3.

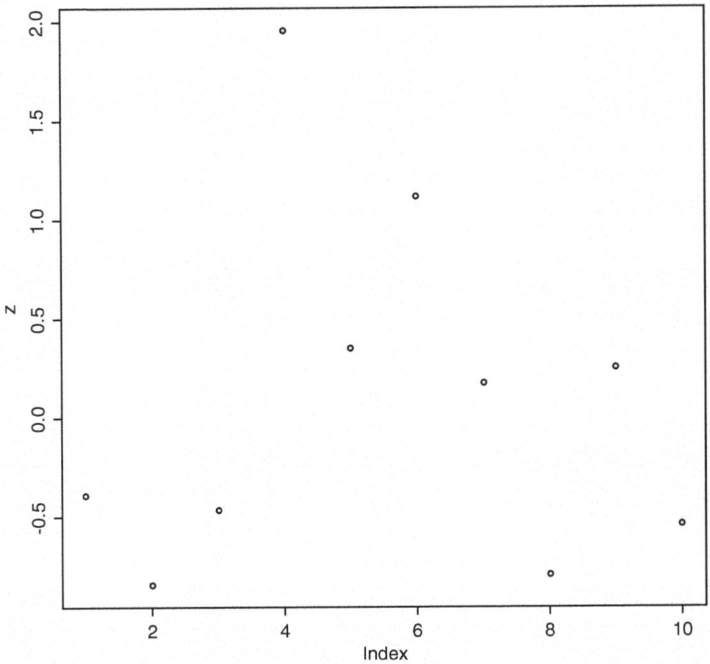

Fig. 2.3 Your first plot

```
z <- rnorm(10)
plot(z)
```

In this case, R simply plots the data in the order they occur in the vector. To plot one variable versus another, just specify the two vectors for the first two arguments. (see Fig. 2.4)

```
x <- -15:15
y <- x^2
plot(x, y)
```

And this is all it takes to generate plots in R, as long as you like the default settings. Of course, the default settings generally will not be sufficient for publication- or presentation-quality graphics. Fortunately, plots in R are very flexible. The table below shows some of the more common arguments to the `plot` function, and some of the common settings. For many more arguments, see the help file for `par` or consult some online materials where http://www.statmethods.net/graphs/ is a useful starting point.

Use of some of the arguments in Table 2.1 is shown in the following example (Fig. 2.5).

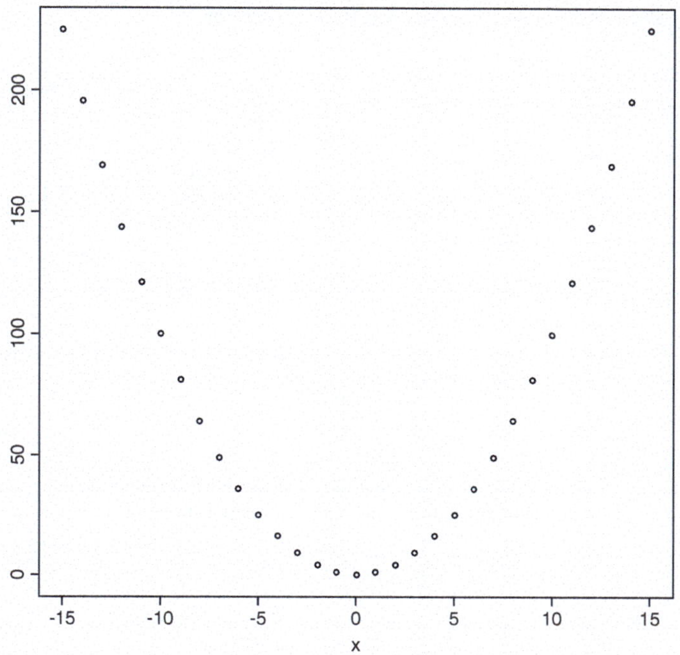

Fig. 2.4 Your second plot

```
plot(x, y, type = "o", xlim = c(-20, 20), ylim = c(-10, 300), pch = 21,
col = "red", bg = "yellow", xlab = "The X variable", ylab = "X squared")
```

The `plot` function is effectively vectorised. It accepts vectors for the first two arguments (which specify the x and y position of your observations), but can also accept vectors for some of the other arguments, including `pch` or `col`. Among other things, this provides an easy way to produce a reference plot demonstrating `R`'s plotting symbols and lines. If you use `R` regularly, you may want to print a copy out (or make your own)—see Fig. 2.6.

```
plot(1:25, rep(1, 25), pch = 1:25, ylim = c(0, 10),
xlab = "", ylab = "",
axes = FALSE) text(1:25, 1.8, as.character(1:25), cex = 0.7)
text(12.5, 2.5, "Default", cex = 0.9)
points(1:25, rep(4, 25), pch = 1:25, col = "blue")
text(1:25, 4.8, as.character(1:25), cex = 0.7, col = "blue")
text(12.5, 5.5, "Blue", cex = 0.9, col = "blue")
points(1:25, rep(7, 25), pch = 1:25, col = "blue", bg = "red")
text(1:25, 7.8, as.character(1:25), cex = 0.7, col = "blue")
text(10, 8.5, "Blue", cex = 0.9, col = "blue")
text(15, 8.5, "Red", cex = 0.9, col = "red")
box()
```

Table 2.1 Some of the more commonly used `plot` arguments

Argument	Common options	Additional information

`col`	`"red"`
	`"blue"`
	`1`
	through
	`657`
		Colour of plotting symbols and lines. Type `colors()` to get list. You can also mix your own colours. See "color specification" in the help file for `par` or http://research.stowers-institute.org/efg/R/Color/Chart/

`bg`	`"red"`
	`"blue"`
	many more
		Colour of fill for some plotting symbols (see below)

`las`	`0`
	`1`
	`2`
	`3`
		Rotation of numeric axis labels

`main`	`Any character string,` `e.g., "plot 1"`	Adds a main title at the top of the plot

`log`	`"x"`
	`"y"`
	`"xy"`
		For making logarithmic scaled axes

`lty`	`0`
	`1` or `"solid"`
	`2` or `"dashed"`
	`3` or `"dashed"`
	through
	`6`
		Line types

`pch`	`0`
	through
	`25`
		Plotting symbols. See below for symbols. Can also use any single character, e.g., "v", or "X" etc

`type`	`"p"` for points `"l"` for line `"b"` for both `"o"` for over `"n"` for none	`"n"` can be handy for setting up a plot that you later add data to
`xlab, ylab`	Any character string, e.g.,"soil depth"	For specifying axis labels

`xlim, ylim`	Any two element vector,
	e.g., `c(0-100)`
	`c(-10-10)`
	`c(55-0)`
		List higher value first to reverse axis

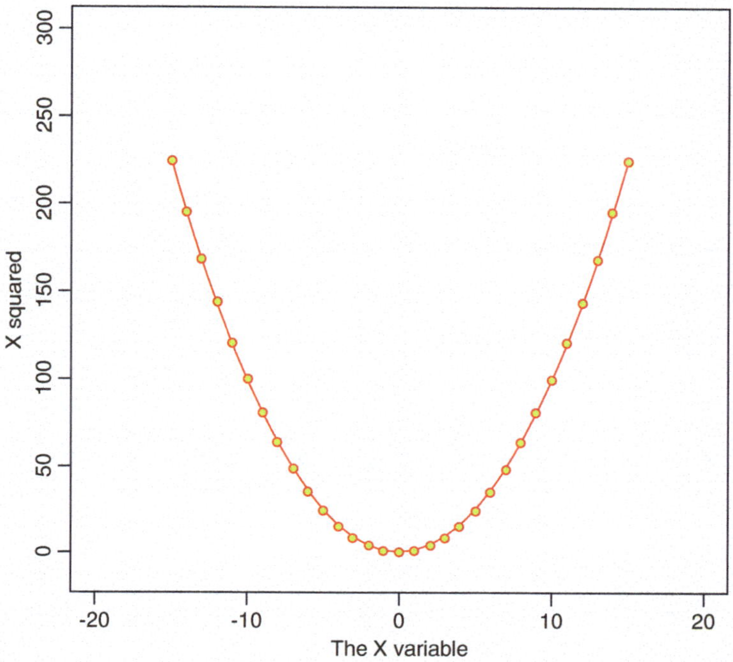

Fig. 2.5 Your first plot using some of the `plot` arguments

2.5.2 *Exercises*

1. Produce a data frame with two columns: x, which ranges from -2π to 2π and has a small interval between values (for plotting), and cosine(x). Plot the cosine(x) vs. x as a line. Repeat, but try some different line types or colours.
2. Read in the data from the `ithir` package called `"USYD_dIndex"`, which contains some observed soil drainage characteristics based on some defined soil colour and drainage index (first column). In the second column is a corresponding prediction which was made by a soil spatial prediction function. Plot the observed drainage index (`DI_observed`) vs. the predicted drainage index (`DI_predicted`). Ensure your plot has appropriate axis limits and labels, and a heading. Try a few plotting symbols and colours. Add some informative text somewhere. If you feel inspired, draw a line of concordance i.e., a 1:1 line on the plot.

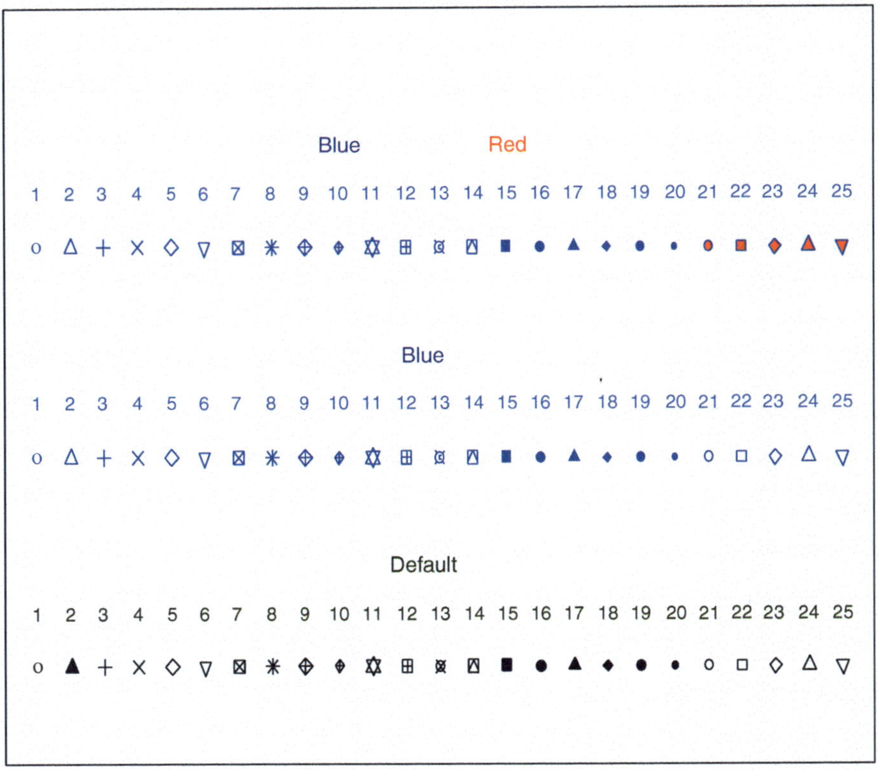

Fig. 2.6 Illustration of some of the `plot` arguments and symbols

2.6 Manipulating Data

2.6.1 Modes, Classes, Attributes, Length, and Coercion

As described before, the mode of an object describes the type of data that it contains. In R, mode is an object attribute. All objects have at least two attributes: mode and length, but may objects have more.

```
x <- 1:10
mode(x)

## [1] "numeric"

length(x)

## [1] 10
```

It is often necessary to change the mode of a data structure, e.g., to have your data displayed differently, or to apply a function that only works with a particular type of data structure. In R this is called coercion. There are many functions in R that have the structure `as.something` that change the mode of a submitted object to "something". For example, say you want to treat numeric data as character data.

```
x <- 1:10
as.character(x)
```

```
##  [1] "1"  "2"  "3"  "4"  "5"  "6"  "7"  "8"  "9"  "10"
```

Or, you may want to turn a matrix into a data frame.

```
X <- matrix(1:30, nrow = 3)
as.data.frame(X)
```

```
##   V1 V2 V3 V4 V5 V6 V7 V8 V9 V10
## 1  1  4  7 10 13 16 19 22 25  28
## 2  2  5  8 11 14 17 20 23 26  29
## 3  3  6  9 12 15 18 21 24 27  30
```

If you are unsure of whether or not a coercion function exists, give it a try—two other common examples are `as.numeric` and `as.vector`.

Attributes are important internally for determining how objects should be handled by various functions. In particular, the `class` attribute determines how a particular object will be handled by a given function. For example, output from a linear regression has the class "`lm`" and will be handled differently by the `print` function than will a data frame, which has the class "`data.frame`". The utility of this object-orientated approach will become more apparent later on.

It is often necessary to know the length of an object. Of course, length can mean different things. Three useful functions for this are `nrow`, `NROW`, and `length`.

The function `nrow` will return the number of rows in a two-dimensional data structure.

```
X <- matrix(1:30, nrow = 3)
X
```

```
##      [,1] [,2] [,3] [,4] [,5] [,6] [,7] [,8] [,9] [,10]
## [1,]    1    4    7   10   13   16   19   22   25    28
## [2,]    2    5    8   11   14   17   20   23   26    29
## [3,]    3    6    9   12   15   18   21   24   27    30
```

```
nrow(X)
```

```
## [1] 3
```

The vertical analog is `ncol`.

```
ncol(X)
```

```
## [1] 10
```

You can get both of these at once with the `dim` function.

```
dim(X)

## [1]  3 10
```

For a vector, use the function `NROW` or `length`.

```
x <- 1:10
NROW(x)

## [1] 10
```

The value returned from the function `length` depends on the type of data structure you submit, but for most data structures, it is the total number of elements.

```
length(X)

## [1] 30

length(x)

## [1] 10
```

2.6.2 Indexing, Sub-setting, Sorting, and Locating Data

Sub-setting and indexing are ways to select specific parts of the data structure (such as specific rows within a data frame) within R. Indexing (also know as sub-scripting) is done using the square braces in R:

```
v1 <- c(5, 1, 3, 8)
v1

## [1] 5 1 3 8

v1[3]

## [1] 3
```

R is very flexible in terms of what can be selected or excluded. For example, the following returns the 1st through 3rd observation:

```
v1[1:3]

## [1] 5 1 3
```

While this returns all but the 4th observation:

```
v1[-4]

## [1] 5 1 3
```

This bracket notation can also be used with relational constraints. For example, if we want only those observations that are <5.0:

```
v1[v1 < 5]

## [1] 1 3
```

This may seem confusing, but if we evaluate each piece separately, it becomes more clear:

```
v1 < 5

## [1] FALSE  TRUE  TRUE FALSE

v1[c(FALSE, TRUE, TRUE, FALSE)]

## [1] 1 3
```

While we are on the topic of subscripts, we should noted that, unlike some other programming languages, the size of a vector in `R` is not limited by its initial assignment. This is true for other data structures as well. To increase the size of a vector, just assign a value to a position that does not currently exist:

```
length(v1)

## [1] 4

v1[8] <- 10
length(v1)

## [1] 8

v1

## [1]  5  1  3  8 NA NA NA 10
```

Indexing can be applied to other data structures in a similar manner as shown above. For data frames and matrices, however, we are now working with two dimensions. In specifying indices, row numbers are given first. We will use our `soil.data` set to illustrate the following few examples:

```
library(ithir)
data(USYD_soil1)
soil.data <- USYD_soil1
dim(soil.data)

## [1] 166  16

str(soil.data)

## 'data.frame': 166 obs. of  16 variables:
##  $ PROFILE       : int  1 1 1 1 1 1 2 2 2 2 ...
##  $ Landclass     : Factor w/ 4 levels "Cropping","Forest",..: 4 4 4 ...
```

```
## $ Upper.Depth : num 0 0.02 0.05 0.1 0.2 0.7 0 0.02 0.05 0.1 ...
## $ Lower.Depth : num 0.02 0.05 0.1 0.2 0.3 0.8 0.02 0.05 0.1 0.2 ...
## $ clay        : int 8 8 8 8 NA 57 9 9 9 NA ...
## $ silt        : int 9 9 10 10 10 8 10 10 10 10 ...
## $ sand        : int 83 83 82 83 79 36 81 80 80 81 ...
## $ pH_CaCl2    : num 6.35 6.34 4.76 4.51 4.64 6.49 5.91 ...
## $ Total_Carbon: num 1.07 0.98 0.73 0.39 0.23 0.35 1.14 ...
## $ EC          : num 0.168 0.137 0.072 0.034 NA 0.059 0.123 ...
## $ ESP         : num 0.3 0.5 0.9 0.2 0.9 0.3 0.3 0.6 NA NA ...
## $ ExchNa      : num 0.01 0.02 0.02 0 0.02 0.04 0.01 0.02 NA NA ...
## $ ExchK       : num 0.71 0.47 0.52 0.38 0.43 0.46 0.7 0.56 NA NA ...
## $ ExchCa      : num 3.17 3.5 1.34 1.03 1.5 9.13 2.92 3.2 NA NA ...
## $ ExchMg      : num 0.59 0.6 0.22 0.22 0.5 5.02 0.51 0.5 NA NA ...
## $ CEC         : num 5.29 3.7 2.86 2.92 2.6 ...
```

If we want to subset out only the first 5 rows, and the first 2 columns:

```
soil.data[1:5, 1:2]
```

```
##   PROFILE      Landclass
## 1       1 native pasture
## 2       1 native pasture
## 3       1 native pasture
## 4       1 native pasture
## 5       1 native pasture
```

If an index is left out, R returns all values in that dimension (you need to include the comma).

```
soil.data[1:2, ]
```

```
##   PROFILE      Landclass Upper.Depth Lower.Depth clay silt sand pH_CaCl2
## 1       1 native pasture        0.00        0.02    8    9   83     6.35
## 2       1 native pasture        0.02        0.05    8    9   83     6.34
##   Total_Carbon    EC ESP ExchNa ExchK ExchCa ExchMg  CEC
## 1         1.07 0.168 0.3   0.01  0.71   3.17   0.59 5.29
## 2         0.98 0.137 0.5   0.02  0.47   3.50   0.60 3.70
```

You can also specify row or column names directly within the brackets—this can be very handy when column order may change in future versions of your code.

```
soil.data[1:5, "Total_Carbon"]
```

```
## [1] 1.07 0.98 0.73 0.39 0.23
```

You can also specify multiple column names using the c function.

```
soil.data[1:5, c("Total_Carbon", "CEC")]
```

```
##   Total_Carbon  CEC
## 1         1.07 5.29
## 2         0.98 3.70
## 3         0.73 2.86
```

```
## 4            0.39 2.92
## 5            0.23 2.60
```

Relational constraints can also be used in indexes. Lets subset out the soil observations that are extremely sodic i.e an ESP greater than 10 %.

```
na.omit(soil.data[soil.data$ESP > 10, ])
```

```
##     PROFILE      Landclass Upper.Depth Lower.Depth clay silt sand pH_CaCl2
## 68       12         Forest        0.02        0.05   22   13   64     4.65
## 111      19 native pasture        0.16        0.26   32   13   56     6.10
## 113      20       Cropping        0.00        0.02    9    9   81     4.64
## 117      20       Cropping        0.30        0.40   37    7   56     6.30
## 118      20       Cropping        0.70        0.80   20   14   67     7.17
## 123      21       Cropping        0.25        0.35   25   16   59     5.05
## 147      26       Cropping        0.15        0.24   51    8   42     6.27
## 148      26       Cropping        0.70        0.80   50    9   40     7.81
##     Total_Carbon    EC  ESP ExchNa ExchK ExchCa ExchMg   CEC
## 68          1.49 0.499 13.0   1.00  0.74   2.17   3.76  6.85
## 111         0.50 0.223 17.4   2.62  0.30   3.74   8.37 11.97
## 113         1.08 0.301 11.1   0.31  0.87   1.01   0.63  3.02
## 117         0.32 0.214 12.1   1.66  0.27   4.19   7.61 12.44
## 118         0.09 0.292 21.2   2.88  0.33   2.86   7.50 10.23
## 123         0.25 0.073 13.2   0.95  0.30   2.00   3.92  5.29
## 147         0.63 0.134 10.4   2.09  0.74   5.09  12.23 14.29
## 148         0.84 0.820 16.4   4.93  0.91   7.52  16.72 23.34
```

While indexing can clearly be used to create a subset of data that meet certain criteria, the `subset` function is often easier and shorter to use for data frames. Sub-setting is used to select a subset of a vector, data frame, or matrix that meets a certain criterion (or criteria). To return what was given in the last example.

```
subset(soil.data, ESP > 10)
```

```
##     PROFILE      Landclass Upper.Depth Lower.Depth clay silt sand pH_CaCl2
## 68       12         Forest        0.02        0.05   22   13   64     4.65
## 111      19 native pasture        0.16        0.26   32   13   56     6.10
## 113      20       Cropping        0.00        0.02    9    9   81     4.64
## 117      20       Cropping        0.30        0.40   37    7   56     6.30
## 118      20       Cropping        0.70        0.80   20   14   67     7.17
## 123      21       Cropping        0.25        0.35   25   16   59     5.05
## 147      26       Cropping        0.15        0.24   51    8   42     6.27
## 148      26       Cropping        0.70        0.80   50    9   40     7.81
##     Total_Carbon    EC  ESP ExchNa ExchK ExchCa ExchMg   CEC
## 68          1.49 0.499 13.0   1.00  0.74   2.17   3.76  6.85
## 111         0.50 0.223 17.4   2.62  0.30   3.74   8.37 11.97
## 113         1.08 0.301 11.1   0.31  0.87   1.01   0.63  3.02
## 117         0.32 0.214 12.1   1.66  0.27   4.19   7.61 12.44
## 118         0.09 0.292 21.2   2.88  0.33   2.86   7.50 10.23
## 123         0.25 0.073 13.2   0.95  0.30   2.00   3.92  5.29
## 147         0.63 0.134 10.4   2.09  0.74   5.09  12.23 14.29
## 148         0.84 0.820 16.4   4.93  0.91   7.52  16.72 23.34
```

Note that the `$` notation does not need to be used in the `subset` function, As with indexing multiple constraints can also be used:

```
subset(soil.data, ESP > 10 & Lower.Depth > 0.3)
```

```
##     PROFILE Landclass Upper.Depth Lower.Depth clay silt sand pH_CaCl2
## 117      20  Cropping        0.30        0.40   37    7   56     6.30
## 118      20  Cropping        0.70        0.80   20   14   67     7.17
## 123      21  Cropping        0.25        0.35   25   16   59     5.05
## 148      26  Cropping        0.70        0.80   50    9   40     7.81
##     Total_Carbon    EC  ESP ExchNa ExchK ExchCa ExchMg   CEC
## 117         0.32 0.214 12.1   1.66  0.27   4.19   7.61 12.44
## 118         0.09 0.292 21.2   2.88  0.33   2.86   7.50 10.23
## 123         0.25 0.073 13.2   0.95  0.30   2.00   3.92  5.29
## 148         0.84 0.820 16.4   4.93  0.91   7.52  16.72 23.34
```

In some cases you may want to select observations that include any one value out of a set of possibilities. Say we only want those observations where `Landclass` is `native pasture` or `forest`. We could use:

```
subset(soil.data, Landclass == "Forest" | Landclass == "native pasture")
```

But, this is an easier way (we are using the `head` function just to limit the number of outputted rows. So try it without the `head` function).

```
head(subset(soil.data, Landclass %in% c("Forest", "native pasture")))
```

```
##   PROFILE      Landclass Upper.Depth Lower.Depth clay silt sand pH_CaCl2
## 1       1 native pasture        0.00        0.02    8    9   83     6.35
## 2       1 native pasture        0.02        0.05    8    9   83     6.34
## 3       1 native pasture        0.05        0.10    8   10   82     4.76
## 4       1 native pasture        0.10        0.20    8   10   83     4.51
## 5       1 native pasture        0.20        0.30   NA   10   79     4.64
## 6       1 native pasture        0.70        0.80   57    8   36     6.49
##   Total_Carbon    EC ESP ExchNa ExchK ExchCa ExchMg   CEC
## 1         1.07 0.168 0.3   0.01  0.71   3.17   0.59  5.29
## 2         0.98 0.137 0.5   0.02  0.47   3.50   0.60  3.70
## 3         0.73 0.072 0.9   0.02  0.52   1.34   0.22  2.86
## 4         0.39 0.034 0.2   0.00  0.38   1.03   0.22  2.92
## 5         0.23    NA 0.9   0.02  0.43   1.50   0.50  2.60
## 6         0.35 0.059 0.3   0.04  0.46   9.13   5.02 14.96
```

Both of the above methods produce the same result, so it just comes down to a matter of efficiency.

Indexing matrices and arrays follows what we have just covered. For example:

```
X <- matrix(1:30, nrow = 3)
X
```

```
##      [,1] [,2] [,3] [,4] [,5] [,6] [,7] [,8] [,9] [,10]
## [1,]    1    4    7   10   13   16   19   22   25    28
## [2,]    2    5    8   11   14   17   20   23   26    29
## [3,]    3    6    9   12   15   18   21   24   27    30
```

```
X[3, 8]
```

```
## [1] 24
```

```
X[, 3]
```

```
## [1] 7 8 9
```

```
Y <- array(1:90, dim = c(3, 10, 3))
Y[3, 1, 1]
```

```
## [1] 3
```

Indexing is a little trickier for lists—you need to use double square braces, `[[i]]`, to specify an element within a list. Of course, if the element within the list has multiple elements, you could use indexing to select specific elements within it.

```
list.1 <- list(1:10, X, Y)
list.1[[1]]
```

```
##  [1]  1  2  3  4  5  6  7  8  9 10
```

It is also possible to use double, triple, etc. indexing with all types of data structures. R evaluates the expression from left to right. As a simple example, lets extract the element on the third row of the second column of the second element of `list.1`:

```
list.1[[2]][3, 2]
```

```
## [1] 6
```

An easy way to divide data into groups is to use the `split` function. This function will divide a data structure (typically a vector or a data frame) into one subset for each level of the variable you would like to split by. The subsets are stored together in a list. Here we split our `soil.data` set into the separate or individual soil profile (splitting by the `PROFILE` column—note output is not shown here for sake of brevity).

```
soil.data.split <- split(soil.data, soil.data$PROFILE)
```

If you apply `split` to individual vectors, the resulting list can be used directly in some plotting or summarizing functions to give you results for each separate group. (There are usually other ways to arrive at this type of result). The `split` function can also be handy for manipulating and analyzing data by some grouping variable, as we will see later.

It is often necessary to sort data. For a single vector, this is done with the function `sort`.

```
x <- rnorm(5)
x
```

```
## [1] -1.1915609  1.3808311  0.9079993  0.2527144 -1.5155076
```

```
y <- sort(x)
y

## [1] -1.5155076 -1.1915609  0.2527144  0.9079993  1.3808311
```

But what if you want to sort an entire data frame by one column? In this case it is necessary to use the function `order`, in combination with indexing.

```
head(soil.data[order(soil.data$clay), ])

##     PROFILE        Landclass Upper.Depth Lower.Depth clay silt sand
## 116      20         Cropping        0.10        0.30    5   11   85
## 1         1   native pasture        0.00        0.02    8    9   83
## 2         1   native pasture        0.02        0.05    8    9   83
## 3         1   native pasture        0.05        0.10    8   10   82
## 4         1   native pasture        0.10        0.20    8   10   83
## 7         2 improved pasture        0.00        0.02    9   10   81
##     pH_CaCl2 Total_Carbon    EC ESP ExchNa ExchK ExchCa ExchMg  CEC
## 116     5.33         0.24 0.033 4.0   0.10  0.18   1.40   0.70 1.90
## 1       6.35         1.07 0.168 0.3   0.01  0.71   3.17   0.59 5.29
## 2       6.34         0.98 0.137 0.5   0.02  0.47   3.50   0.60 3.70
## 3       4.76         0.73 0.072 0.9   0.02  0.52   1.34   0.22 2.86
## 4       4.51         0.39 0.034 0.2   0.00  0.38   1.03   0.22 2.92
## 7       5.91         1.14 0.123 0.3   0.01  0.70   2.92   0.51 3.59
```

The function `order` returns a vector that contains the row positions of the ranked data:

```
order(soil.data$clay)

##   [1] 116   1   2   3   4   7   8   9  16 113 114 115  14  31  32  61  62
##  [18]  64  45 126 149 150  37  38  39 101 102 103 107 108 151  34  42  44
##  [35]  77  13  43  46  50 109 125  78 144  35  51  79  96 110 119 120 121
##  [52] 127 134  67  97 130 135 145 161  36  98 136 140 146 160 162  17  52
##  [69]  83 104 118 131 139 141  65  84  85  68  69  89  99  53 156  72 123
##  [86] 137  54  56  80  95 105  57  90  74 132  58  81  91 128  20  55 111
## [103] 142  86 163 106 154  11 129  18  87 112 117  25  75  21 155  66  92
## [120]  22  26 152 164 133 138  47  71  41  59  93  60 148  24 147  12 158
## [137] 165   6  82 166  48  88 159  28  29  94  76 100  40   5  10  15  19
## [154]  23  27  30  33  49  63  70  73 122 124 143 153 157
```

The previous discussion in this section showed how to isolate data that meet certain criteria from a data structure. But sometimes it is important to know where data resides in its original data structure. But sometimes it is important to know where data resides in its original data structure. To functions that are handy for locating data within an `R` data structure are `match` and `which`. The `match` function will tell you where specific values reside in a data structure, while the `which` function will return the locations of values that meet certain criteria.

```
match(c(25.85, 11.45, 9.23), soil.data$CEC)

## [1] 41 59 18
```

and to check the result...

```
soil.data[c(41, 59, 18), ]
```

```
##    PROFILE         Landclass Upper.Depth Lower.Depth clay silt sand
## 41       7          Cropping         0.7         0.8   47   11   42
## 59      10 improved pasture          0.1         0.2   47   12   42
## 18       3            Forest         0.7         0.8   37    9   54
##    pH_CaCl2 Total_Carbon    EC ESP ExchNa ExchK ExchCa ExchMg   CEC
## 41     6.70         0.23 0.063 2.2   0.61  0.53  14.22  12.95 25.85
## 59     5.94         0.70 0.039 0.1   0.01  0.74   6.76   2.61 11.45
## 18     6.05         0.17 0.088 0.7   0.06  0.33   6.15   2.35  9.23
```

Note that the match function matches the first observation only (this makes it difficult to use when there are multiple observations of the same value). This function is vectorised. The match function is useful for finding the location of the unique values, such as the maximum.

```
match(max(soil.data$CEC, na.rm = TRUE), soil.data$CEC)
```

```
## [1] 95
```

Note the call to the na.rm argument in the max function as a means to overlook the presence of NA values. So what is the maximum CEC value in our soil.data set.

```
soil.data$CEC[95]
```

```
## [1] 28.21
```

The which function, on the other hand, will return all locations that meet the criteria.

```
which(soil.data$ESP > 5)
```

```
##  [1]  68 101 102 103 104 107 108 109 111 113 114 117 118 123 146 147 148
## [18] 159
```

Of course, you can specify multiple constraints.

```
which(soil.data$ESP > 5 & soil.data$clay > 30)
```

```
## [1] 111 117 147 148 159
```

The which function can also be useful for locating missing values.

```
which(is.na(soil.data$ESP))
```

```
##  [1]   9  10  15  21  24  45  49  58  61  63  80  81  89 110 112 115 121
## [18] 129 138 150
```

```
soil.data$ESP[c(which(is.na(soil.data$ESP)))]
```

```
##  [1] NA NA NA NA NA NA NA NA NA NA NA NA NA NA NA NA NA NA NA NA
```

2.6.3 Factors

For many analyses, it is important to distinguish between quantitative (i.e., continuous) and categorical (i.e., discrete) variables. Categorical data are called factors in R. Internally, factors are stored as numeric data (as a check with mode will tell you), but they are handled as categorical data in statistical analyses. Factors are a class of data in R. R automatically recognizes non-numerical data as factors when the data are read in, but if numerical data are to be used as a factor (or if character data are generated within R and not read in, conversion to a factor must be specified explicitly. In R, the function factor does this.

```
a <- c(rep(0, 4), rep(1, 4))
a

## [1] 0 0 0 0 1 1 1 1

a <- factor(a)
a

## [1] 0 0 0 0 1 1 1 1
## Levels: 0 1
```

The levels that R assigns to your factor are by default the unique values given in your original vector. This is often fine, but you may want to assign more meaningful levels. For example, say you have a vector that contains soil drainage class categories.

```
soil.drainage<- c("well drained", "imperfectly drained", "poorly drained",
"poorly drained", "well drained", "poorly drained")
```

If you designate this as a factor, the default levels will be sorted alphabetically.

```
soil.drainage1 <- factor(soil.drainage)
soil.drainage1

## [1] well drained        imperfectly drained poorly drained
## [4] poorly drained      well drained        poorly drained
## Levels: imperfectly drained poorly drained well drained

as.numeric(soil.drainage1)

## [1] 3 1 2 2 3 2
```

If you specify levels as an argument of the factor function, you can control the order of the levels.

```
soil.drainage2 <- factor(soil.drainage, levels = c("well drained",
"imperfectly drained", "poorly drained"))
as.numeric(soil.drainage2)

## [1] 1 2 3 3 1 3
```

This can be useful for obtaining a logical order in statistical output or summaries.

2.6.4 Combining Data

Data frames (or vectors or matrices) often need to be combined for analysis or plotting. Three `R` functions that are very useful for combining data are `rbind` and `cbind`. The function `rbind` simply "stacks" objects on top of each other to make a new object ("row bind"). The function `cbind` ("column bind") carries out an analogous operation with columns of data.

```
soil.info1 <- data.frame(soil = c("Vertosol", "Hydrosol", "Sodosol"),
response = 1:3)
soil.info1

##       soil response
## 1 Vertosol        1
## 2 Hydrosol        2
## 3  Sodosol        3

soil.info2 <- data.frame(soil = c("Chromosol", "Dermosol", "Tenosol"),
response = 4:6)
soil.info2

##        soil response
## 1 Chromosol        4
## 2  Dermosol        5
## 3   Tenosol        6

soil.info <- rbind(soil.info1, soil.info2)
soil.info

##        soil response
## 1  Vertosol        1
## 2  Hydrosol        2
## 3   Sodosol        3
## 4 Chromosol        4
## 5  Dermosol        5
## 6   Tenosol        6

a.column <- c(2.5, 3.2, 1.2, 2.1, 2, 0.5)
soil.info3 <- cbind(soil.info, SOC = a.column)
soil.info3

##        soil response SOC
## 1  Vertosol        1 2.5
## 2  Hydrosol        2 3.2
## 3   Sodosol        3 1.2
## 4 Chromosol        4 2.1
## 5  Dermosol        5 2.0
## 6   Tenosol        6 0.5
```

2.6.5 *Exercises*

1. Using the `soil.data` set, return the following:
   (a) The first 10 rows of the columns `clay`, `Total_Carbon`, and `ExchK`
   (b) The column `CEC` for the `Forest` land class
   (c) Use the `subset` function to create a new data frame that has only data for the cropping land class
   (d) How many soil profiles are there? Actually write some script to determine this rather than look at the data frame
2. Using the same data set, find the location and value of the maximum, minimum and median soil carbon (`Total_Carbon`)
3. Make a new data frame which is sorted by the upper soil depth (`Upper.Depth`). Can you sort it in decreasing order (Hint: Check the help file)
4. Make a new data frame which contains the columns `PROFILE`, `Landclass`, `Upper.Depth`, and `Lower.Depth`. Make another data frame which contains just the information regarding the exchangeable cations e.g., `ExchNa`, `ExchK`, `ExchCa`, and `ExchMg`. Make a new data frame which combines these two separate data frames together
5. Make separate data frame for each of the land classes. Now make a new data frame which combines the four separate data frames together.

2.7 Exploratory Data Analysis

2.7.1 *Summary Statistics*

We will again use the `soil.data` set to demonstrate calculation of summary statistics. Just for recall, lets see what is in this data frame.

```
library(ithir)
data(USYD_soil1)
soil.data <- USYD_soil1
names(soil.data)
```

```
##  [1] "PROFILE"      "Landclass"    "Upper.Depth"  "Lower.Depth"
##  [5] "clay"         "silt"         "sand"         "pH_CaCl2"
##  [9] "Total_Carbon" "EC"           "ESP"          "ExchNa"
## [13] "ExchK"        "ExchCa"       "ExchMg"       "CEC"
```

Here are some useful functions (and note the usage of the `na.rm` argument) for calculation of means (`mean`), medians (`median`), standard deviations (`sd`) and variances (`var`):

```
mean(soil.data$clay, na.rm = TRUE)
```

```
## [1] 26.95302
```

```
median(soil.data$clay, na.rm = TRUE)

## [1] 21

sd(soil.data$clay, na.rm = TRUE)

## [1] 15.6996

var(soil.data$clay, na.rm = TRUE)

## [1] 246.4775
```

R has a built-in function for summarizing vectors or data frames called `summary`. This function is a generic function—what it returns is dependent on the type of data set to it. Applying the `summary` function to the first 6 columns in the `soil.data` set results in the following output:

```
summary(soil.data[, 1:6])

##     PROFILE                    Landclass   Upper.Depth      Lower.Depth
##  Min.   : 1.00   Cropping          :49   Min.   :0.0000   Min.   :0.0200
##  1st Qu.: 8.00   Forest            :50   1st Qu.:0.0200   1st Qu.:0.0500
##  Median :15.00   improved pasture:35   Median :0.0500   Median :0.1000
##  Mean   :14.73   native pasture   :32   Mean   :0.1816   Mean   :0.2464
##  3rd Qu.:22.00                           3rd Qu.:0.2000   3rd Qu.:0.3000
##  Max.   :29.00                           Max.   :0.7000   Max.   :0.8000
##
##       clay             silt
##  Min.   : 5.00   Min.   : 6.0
##  1st Qu.:15.00   1st Qu.:11.0
##  Median :21.00   Median :15.0
##  Mean   :26.95   Mean   :16.5
##  3rd Qu.:37.00   3rd Qu.:20.0
##  Max.   :68.00   Max.   :32.0
##  NA's   :17      NA's   :1
```

Notice the difference between numerical and categorical variables. The `summary` function should probably be your first stop after organizing your data, and before analyzing it—it provides an easy way to check for wildly erroneous values.

2.7.2 Histograms and Box Plots

Box plots and histograms are simple but useful ways of summarizing data. You can generate a histogram in R using the function `hist`.

```
hist(soil.data$clay)
```

The histogram (Fig. 2.7) can be made to look nicer, by applying some of the plotting parameters or arguments that we covered for the `plot` function. There are

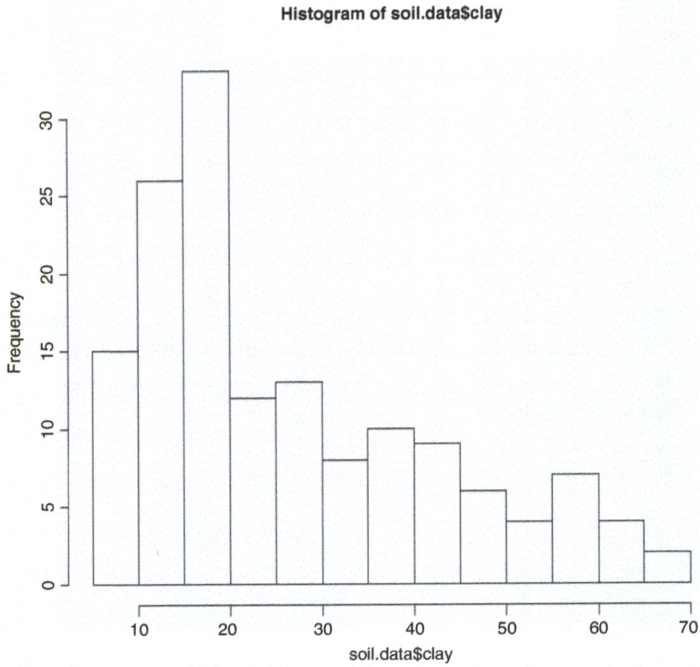

Fig. 2.7 Histogram of clay content from `soil.data`

also some additional "plotting" arguments that can be sourced in the `hist` help file. One of these is the ability to specify the number or location of breaks in the histogram.

Box plots are also a useful way to summarize data. We can use it simply, for example, summarize the clay content in the `soil.data` (Fig. 2.8).

```
boxplot(soil.data$clay)
```

By default, the heavy line shows the median, the box shows the 25th and 75th percentiles, the "whiskers" show the extreme values, and points show outliers beyond these.

Another approach is to plot a single variable by some factor. Here we will plot `Total_Carbon` by `Landclass` (Fig. 2.9).

```
boxplot(Total_Carbon ~ Landclass, data = soil.data)
```

Note the use of the tilde symbol "∼" in the above command. The code `Total_Carbon∼Landclass` is analogous to a model formula in this case, and simply indicates that `Total_Carbon` is described by `Landclass` and should be split up based on the category of this variable. We will see more of this character with the specification of soil spatial prediction functions later on.

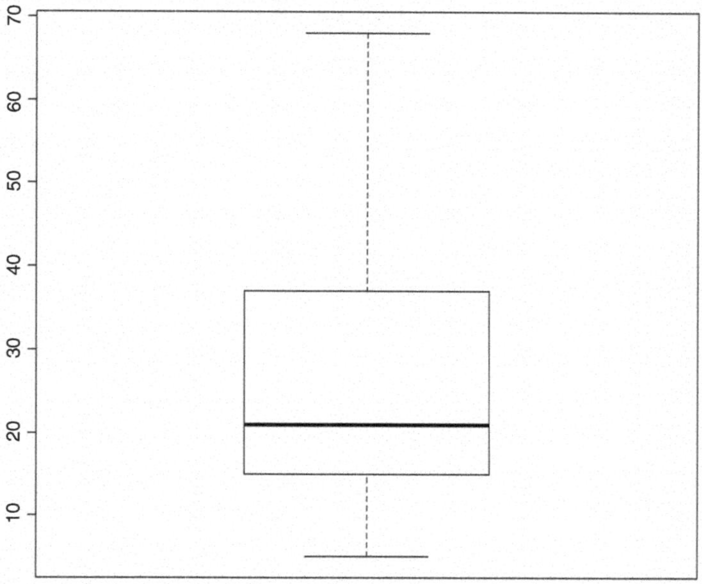

Fig. 2.8 Boxplot of clay content from `soil.data`

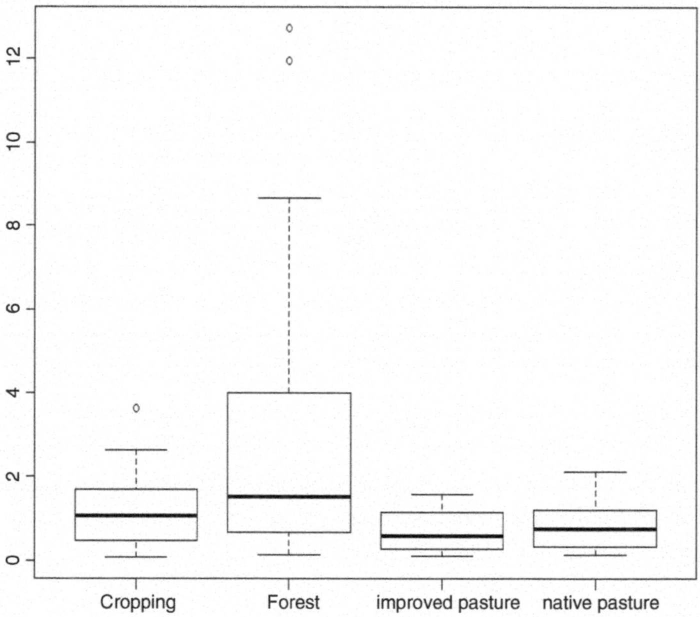

Fig. 2.9 Box plot of total carbon with respect to landclass

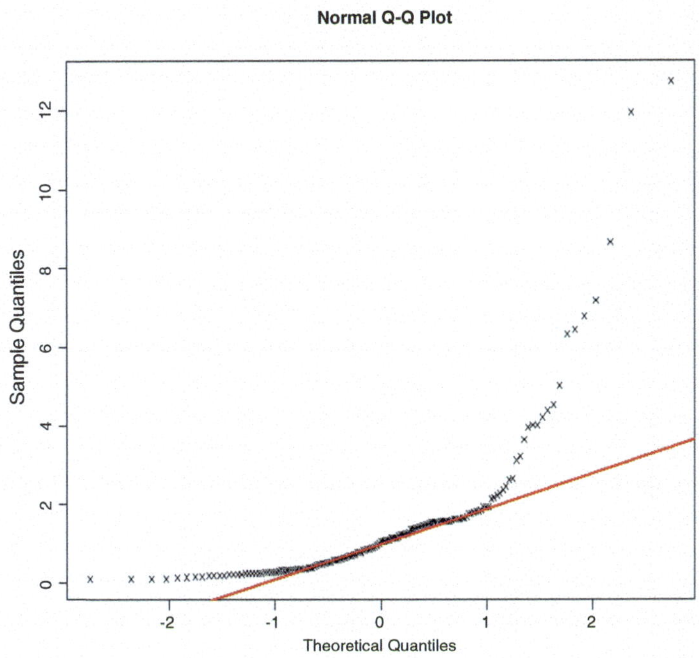

Fig. 2.10 QQ plot of total carbon in the `soil.data` set

2.7.3 Normal Quantile and Cumulative Probability Plots

One way to assess the normality of the distribution of a given variable is with a quantile-quantile plot. This plot shows data values vs. quantiles based on a normal distribution (Fig. 2.10).

```
qqnorm(soil.data$Total_Carbon, plot.it = TRUE, pch = 4, cex = 0.7)
qqline(soil.data$Total_Carbon, col = "red", lwd = 2)
```

There definitely seems to be some deviation from normality here. This is not unusual for soil carbon information. It is common (in order to proceed with statistical modelling) to perform a transformation of sorts in order to get these data to conform to a normal distribution—lets see if a log transformation works any better (Fig. 2.11).

```
qqnorm(log(soil.data$Total_Carbon), plot.it = TRUE, pch = 4, cex = 0.7)
qqline(log(soil.data$Total_Carbon), col = "red", lwd = 2)
```

Finally, another useful data exploratory tool is quantile calculations. R will return the quantiles of a given data set with the `quantile` function. Note that there are

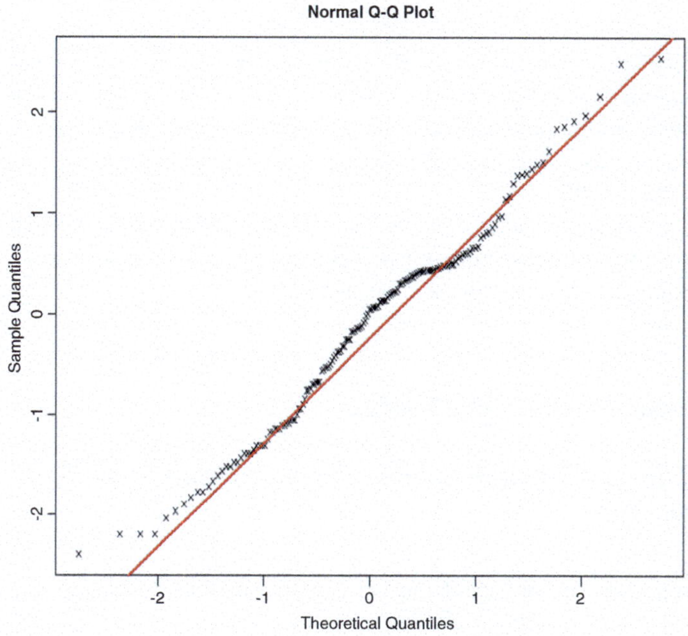

Fig. 2.11 QQ plot of log-transformed total carbon in the `soil.data` set

nine different algorithms available for doing this—you can find descriptions in the help file for `quantile`.

```
quantile(soil.data$Total_Carbon, na.rm = TRUE)

##    0%   25%   50%   75%  100%
##  0.09  0.39  1.05  1.60 12.74

quantile(soil.data$Total_Carbon, na.rm = TRUE, probs = seq(0, 1, 0.05))

##     0%      5%     10%     15%     20%     25%     30%     35%     40%     45%
##  0.090   0.170   0.230   0.270   0.328   0.390   0.502   0.604   0.730   0.866
##    50%     55%     60%     65%     70%     75%     80%     85%     90%     95%
##  1.050   1.150   1.268   1.448   1.548   1.600   1.762   2.026   2.928   4.494
##   100%
## 12.740

quantile(soil.data$Total_Carbon, na.rm = TRUE, probs = seq(0.9, 1, 0.01))

##      90%      91%      92%      93%      94%      95%      96%      97%      98%
##   2.9280   3.3208   3.9128   4.0152   4.2388   4.4940   5.5920   6.4488   7.0536
##      99%     100%
##   9.8344  12.7400
```

2.7.4 Exercises

1. Using the `soil.data` set firstly determine the summary statistics for each of the numerical or quantitative variables. You want to calculate things like maximum, minimum, mean, median, standard deviation, and variance. There are a couple of ways to do this. However, put all the results into a data frame and export as a text file.
2. Generate histograms and QQ plots for each of the quantitative variables. Do any need some sort of transformation so that their distribution is normal. If so, do the transformation and perform the plots again.

2.8 Linear Models: The Basics

2.8.1 The `lm` Function, Model Formulas, and Statistical Output

In R several classical statistical models can be implemented using one function: `lm` (for linear model). The `lm` function can be used for simple and multiple linear regression, analysis of variance (ANOVA), and analysis of covariance (ANCOVA). The help file for `lm` lists the following.

```
lm(formula, data, subset, weights, na.action, method
= "qr", model = TRUE, x = FALSE, y = FALSE, qr = TRUE,
singular.ok = TRUE, contrasts = NULL, offset, ...)
```

The first argument in the `lm` function call (`formula`) is where you specify the structure of the statistical model. This approach is used in other R functions as well, such as `glm`, `gam`, and others, which we will later investigate. The most common structures of a statistical model are:

y ~ x	Simple linear regression of y on x
y ~ x + z	Multiple regression of y on x and z

There are many others, but it is these model structures that are used in one form or another for DSM.

There is some similarity between the statistical output in R and in other statistical software programs. However, by default, R usually gives only basic output. More detailed output can be retrieved with the `summary` function. For specific statistics, you can use "extractor" functions, such as `coef` or `deviance`. Output from the `lm` function is of the class `lm`, and both default output and specialized output from extractor functions can be assigned to objects (this is of course true for other model objects as well). This quality is very handy when writing code that uses the results of statistical models in further calculations or in compiling summaries.

2.8.2 Linear Regression

To demonstrate simple linear regression in R, we will again use the soil.data set. Here we will regress CEC content on clay.

```
summary(cbind(clay = soil.data$clay, CEC = soil.data$CEC))

##        clay              CEC
##  Min.   : 5.00    Min.   : 1.900
##  1st Qu.:15.00    1st Qu.: 5.350
##  Median :21.00    Median : 8.600
##  Mean   :26.95    Mean   : 9.515
##  3rd Qu.:37.00    3rd Qu.:12.110
##  Max.   :68.00    Max.   :28.210
##  NA's   :17       NA's   :5
```

The (alternative) hypothesis here is that clay content is a good predictor of CEC. As a start, let us have a look at what the data looks like (Fig. 2.12).

```
plot(soil.data$clay, soil.data$CEC)
```

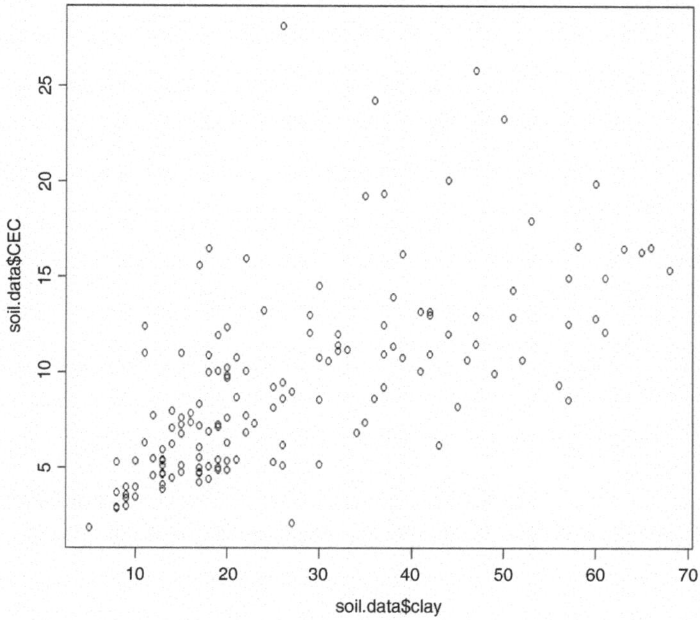

Fig. 2.12 plot of CEC against clay from the soil.data set

There appears to be some meaningful relationship. To fit a linear model, we can use the `lm` function:

```
mod.1 <- lm(CEC ~ clay, data = soil.data, y = TRUE, x = TRUE)
mod.1

##
## Call:
## lm(formula = CEC ~ clay, data = soil.data, x = TRUE, y = TRUE)
##
## Coefficients:
## (Intercept)         clay
##      3.7791       0.2053
```

R returns only the call and coefficients by default. You can get more information using the `summary` function.

```
summary(mod.1)

##
## Call:
## lm(formula = CEC ~ clay, data = soil.data, x = TRUE, y = TRUE)
##
## Residuals:
##     Min      1Q  Median      3Q     Max
## -7.1829 -2.3369 -0.6767  1.0185 19.0924
##
## Coefficients:
##             Estimate Std. Error t value Pr(>|t|)
## (Intercept)  3.77913    0.63060   5.993 1.58e-08 ***
## clay         0.20533    0.02005  10.240  < 2e-16 ***
## ---
## Signif. codes:  0 '***' 0.001 '**' 0.01 '*' 0.05 '.' 0.1 ' ' 1
##
## Residual standard error: 3.783 on 144 degrees of freedom
##   (20 observations deleted due to missingness)
## Multiple R-squared:  0.4214,Adjusted R-squared:  0.4174
## F-statistic: 104.9 on 1 and 144 DF,  p-value: < 2.2e-16
```

Clay does appear to be a significant predictor of CEC as one would generally surmise using some basic soil science knowledge. As mentioned above, the output from the `lm` function is an object of class `lm`. These objects are lists that contain at least the following elements (you can find this list in the help file for `lm`):

`coefficients`	a named vector of model coefficients
`residuals`	model residuals (observed - predicted values)
`fitted.value`	the fitted mean values
`rank`	a numeric rank of the fitted linear model
`weights`	(only for weighted fits) the specified weights
`df.residual`	the residual degrees of freedom
`call`	the matched call

`terms` the terms object used
`contrasts` (only where relevant) the contrasts used
`xlevels` (only where relevant) a record of the levels of the factors used in fitting
`offset` the offset used (missing if none were used)
`y` if requested, the response used
`x` if requested, the model matrix used
`model` if requested (the default), the model frame used
`na.action` (where relevant) information returned by model frame on the handling of NAs

```
class(mod.1)

## [1] "lm"
```

To get at the elements listed above, you can simply index the `lm` object, i.e., call up part of the list.

```
mod.1$coefficients

## (Intercept)        clay
##   3.7791256   0.2053256
```

However, `R` has several extractor functions designed precisely for pulling data out of statistical model output. Some of the most commonly used ones are: `add1`, `alias`, `anova`, `coef`, `deviance`, `drop1`, `effects`, `family`, `formula`, `kappa`, `labels`, `plot`, `predict`, `print`, `proj`, `residuals`, `step`, `summary`, and `vcov`. For example:

```
coef(mod.1)

## (Intercept)        clay
##   3.7791256   0.2053256

head(residuals(mod.1))

##          1          2          3          4          6          7
## -0.1317300 -1.7217300 -2.5617300 -2.5017300 -0.5226822 -2.0370556
```

As mentioned before, the `summary` function is a generic function—what it does and what it returns is dependent on the class of its first argument. Here is a list of what is available from the `summary` function for this model:

```
names(summary(mod.1))

##  [1] "call"          "terms"         "residuals"     "coefficients"
##  [5] "aliased"       "sigma"         "df"            "r.squared"
##  [9] "adj.r.squared" "fstatistic"    "cov.unscaled"  "na.action"
```

To extract some of the information in the `summary` which is of a list structure, we would use:

```
summary(mod.1)[[4]]

##              Estimate Std. Error   t value     Pr(>|t|)
## (Intercept) 3.7791256 0.63060292  5.992877 1.577664e-08
## clay        0.2053256 0.02005048 10.240431 7.910414e-19
```

To look at the statistical summary. What is the R^2 of `mod.1`?

```
summary(mod.1)[["r.squared"]]

## [1] 0.4213764
```

This flexibility is useful, but makes for some redundancy in R. For many model statistics, there are three ways to get your data: an extractor function (such as `coef`), indexing the `lm` object, and indexing the `summary` function. The best approach is to use an extractor function whenever you can. In some cases, the `summary` function will return results that you can not get by indexing or using extractor functions.

Once we have fit a model in R, we can generate predicted values using the `predict` function.

```
head(predict(mod.1))

##         1         2         3         4         6         7
##  5.421730  5.421730  5.421730  5.421730 15.482682  5.627056
```

Lets plot the observed vs. the predict from this model (Fig. 2.13).

```
plot(mod.1$y, mod.1$fitted.values)
```

As we will see later on, the `predict` function works for a whole range of statistical models in R—not just `lm` objects. We can treat the predictions as we would any vector. For example we can add them to the above plot or put them back in the original data frame. The `predict` function can also give confidence and prediction intervals.

```
head(predict(mod.1, int = "conf"))

##         fit       lwr       upr
## 1  5.421730  4.437845  6.405615
## 2  5.421730  4.437845  6.405615
## 3  5.421730  4.437845  6.405615
## 4  5.421730  4.437845  6.405615
## 6 15.482682 14.152940 16.812425
## 7  5.627056  4.673657  6.580454
```

A quick way to demonstrate multiple linear regression in R, we will regress `CEC` on `clay` plus the exchangeable cations: `ExchNa` and `ExchCa`. First lets subset these data out, then get their summary statistics.

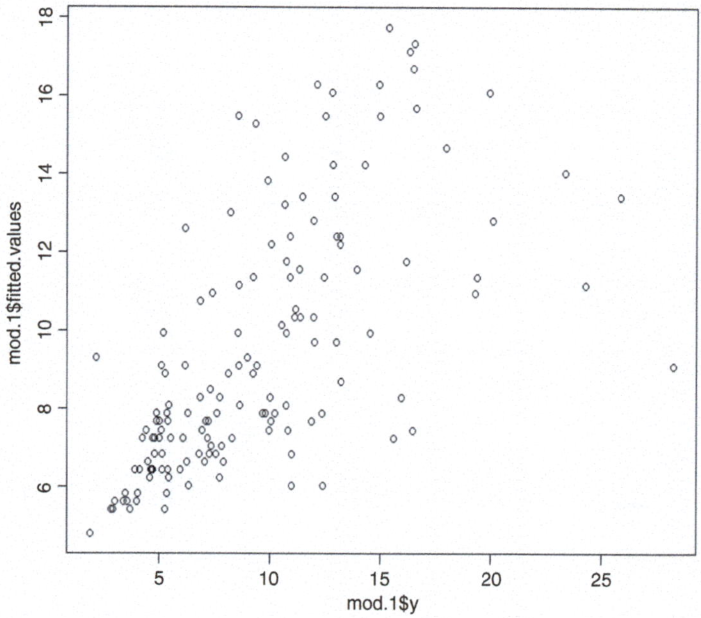

Fig. 2.13 observed vs. predicted plot of CEC from `soil.data`

```
subs.soil.data <- soil.data[, c("clay", "CEC", "ExchNa", "ExchCa")]
summary(subs.soil.data)
```

```
##      clay            CEC             ExchNa            ExchCa
##  Min.    : 5.00   Min.   : 1.900   Min.   :0.0000   Min.   : 1.010
##  1st Qu.:15.00   1st Qu.: 5.350   1st Qu.:0.0200   1st Qu.: 2.920
##  Median :21.00   Median : 8.600   Median :0.0500   Median : 4.610
##  Mean   :26.95   Mean   : 9.515   Mean   :0.2563   Mean   : 5.353
##  3rd Qu.:37.00   3rd Qu.:12.110   3rd Qu.:0.1500   3rd Qu.: 7.240
##  Max.   :68.00   Max.   :28.210   Max.   :6.8100   Max.   :25.390
##  NA's   :17      NA's   :5        NA's   :5        NA's   :5
```

A quick way to look for relationships between variables in a data frame is with the `cor` function. Note the use of the `na.omit` function.

```
cor(na.omit(subs.soil.data))
```

```
##             clay       CEC      ExchNa      ExchCa
## clay   1.0000000 0.6491351  0.17104962  0.54615778
## CEC    0.6491351 1.0000000  0.33369789  0.87570029
## ExchNa 0.1710496 0.3336979  1.00000000 -0.02055496
## ExchCa 0.5461578 0.8757003 -0.02055496  1.00000000
```

To visualize these relationships, we can use `pairs` (Fig. 2.14).

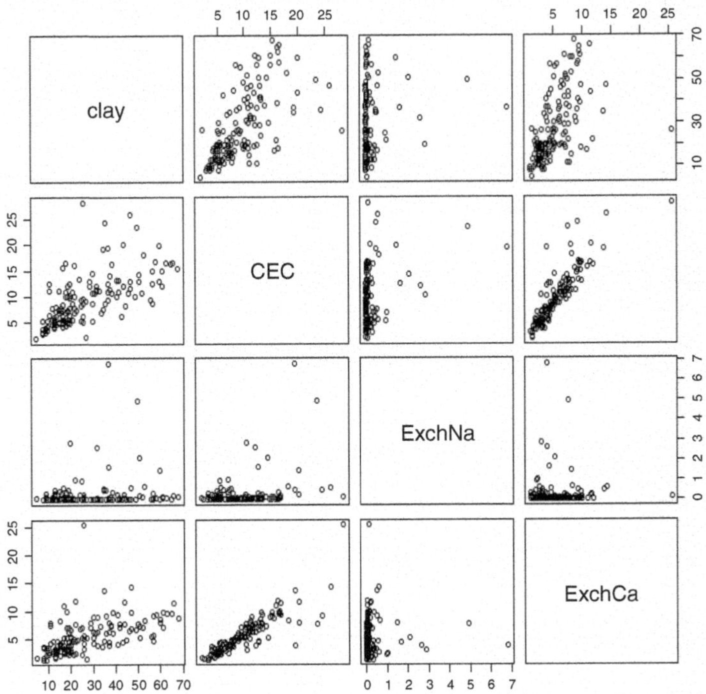

Fig. 2.14 A pairs plot of a select few soil attributes from the `soil.data` set

```
pairs(na.omit(subs.soil.data))
```

There are some interesting relationships here. Now for fitting the model:

```
mod.2 <- lm(CEC ~ clay + ExchNa + ExchCa, data = subs.soil.data)
summary(mod.2)

##
## Call:
## lm(formula = CEC ~ clay + ExchNa + ExchCa, data =
     subs.soil.data)
##
## Residuals:
##     Min      1Q  Median      3Q     Max
## -5.2008 -0.7065 -0.0470  0.6455  9.4025
##
## Coefficients:
##             Estimate Std. Error t value Pr(>|t|)
## (Intercept) 1.048318   0.274264   3.822 0.000197 ***
## clay        0.050503   0.009867   5.119 9.83e-07 ***
## ExchNa      2.018149   0.163436  12.348  < 2e-16 ***
## ExchCa      1.214156   0.046940  25.866  < 2e-16 ***
```

```
## ---
## Signif. codes:  0 '***' 0.001 '**' 0.01 '*' 0.05 '.' 0.1 ' ' 1
##
## Residual standard error: 1.522 on 142 degrees of freedom
##   (20 observations deleted due to missingness)
## Multiple R-squared:  0.9076,Adjusted R-squared:  0.9057
## F-statistic: 465.1 on 3 and 142 DF,  p-value: < 2.2e-16
```

For much of the remainder of this book, we will be investigating these regression type relationships using a variety of different model types for soil spatial prediction. These fundamental modelling concepts of the `lm` will become useful as we progress ahead.

2.8.3 Exercises

1. Using the `soil.data` set firstly generate a correlation matrix of all the soil variables.
2. Choose two variables that you think would be good to regress against each other, and fit a model. Is the model any good? Can you plot the observed vs. predicted values? Can you draw a line of concordance (maybe want to consult an appropriate help file to do this). Something a bit trickier: Can you add the predictions to the data frame `soil.data` correctly.
3. Now repeat what you did for the previous question, except this time perform a multiple linear regression i.e., use more than 1 predictor variable to make a prediction of the variable you are targeting.

2.9 Advanced Work: Developing Algorithms with R

One of the advantages of using a scripting language such as R is that we can develop algorithms or script a set of commands for problem-solving. We illustrate this by developing an algorithm for generating a catena, or more correctly, a digital toposequence from a digital elevation model (DEM). To make it more interesting, we will not use any function from external libraries other than basic R functions. Before scripting, it is best to write out the algorithm or sequence of routines to use.

A toposequence can be described as a transect (not necessarily a straight line) which begins at a hilltop and ends at a valley bottom or a stream (Odgers et al. 2008). To generate a principal toposequence, one can start from the highest point in an area. If we numerically consider a rainfall as a discrete packet of "precipitation" that falls on the highest elevation pixel, then this precipitation will mostly move to its neighbor with the highest slope. As DEM is stored in matrix format, we simply

Fig. 2.15 Indexation of 3 × 3 neighborhood

[-1,1]	[0,1]	[1,1]
[-1,0]	[0,0] Current cell	[1,0]
[-1,-1]	[0,-1]	[1,-1]

look around its 3 × 3 neighbors and determine which pixel is the lowest elevation. The algorithm for principal toposequence can be written as:

1. Determine the highest point in an area.
2. Determine its 3 × 3 neighbor, and determine whether there are lower points?
3. If yes, set the lowest point as the next point in the toposequence, and then repeat step 2. If no, the toposequence has ended.

To facilitate the 3 × 3 neighbor search in R, we can code the neighbors using its relative coordinates. If the current cell is designated as [0, 0], then its left neighbor is [−1, 0], and so on. We can visualize it as follows in Fig. 2.15.

If we designate the current cell [0, 0] as *z1*, the function below will look for the lowest neighbor for pixel z1 in a DEM.

```
# function to find the lowest 3 x 3 neighbor
find_steepest <- function(dem, row_z, col_z)
{
    z1 = dem[row_z, col_z]  #elevation
    # return the elevation of the neighboring values
    dir = c(-1, 0, 1)  #neighborhood index
    nr = nrow(dem)
    nc = ncol(dem)
    pz = matrix(data = NA, nrow = 3, ncol = 3) #placeholder
         for the values
    for (i in 1:3) {
        for (j in 1:3) {
            if (i != 0 & j != 0) {
                ro <- row_z + dir[i]
                co <- col_z + dir[j]
                if (ro > 0 & co > 0 & ro < nr & co < nc) {
                  pz[i, j] = dem[ro, co]
                }
            }
```

```
        }
    }

    pz <- pz - z1  # difference of neighbors from centre value
    # find lowest value
    min_pz <- which(pz == min(pz, na.rm = TRUE), arr.ind = TRUE)
    row_min <- row_z + dir[min_pz[1]]
    col_min <- col_z + dir[min_pz[2]]
    retval <- c(row_min, col_min, min(pz, na.rm = TRUE))
    return(retval)  #return the minimum
}
```

The principal toposequence code can be implemented as follows. First we load in a small dataset called `topo_dem` from the `ithir` package.

```
library(ithir)
data(topo_dem)
str(topo_dem)
```

```
##  num [1:109, 1:110] 121 121 120 118 116 ...
##  - attr(*, "dimnames")=List of 2
##   ..$ : NULL
##   ..$ : chr [1:110] "V1" "V2" "V3" "V4" ...
```

Now we want to create a data matrix to store the result of the toposequence i.e. the row, column, and elevation values that are selected using the `find_steepest` function.

```
transect <- matrix(data = NA, nrow = 20, ncol = 3)
```

Now we want to find within that matrix that maximum elevation value and its corresponding row and column position.

```
max_elev <- which(topo_dem == max(topo_dem), arr.ind = TRUE)
row_z = max_elev[1]  # row of max_elev
col_z = max_elev[2]  # col of max_elev
z1 = topo_dem[row_z, col_z]  # max elevation

# Put values into the first entry of the transect object
t <- 1
transect[t, 1] = row_z
transect[t, 2] = col_z
transect[t, 3] = z1
lowest = FALSE
```

Below we use the `find_steepest` function. It is embedded into a `while` conditional loop, such that the routine will run until neither of the surrounding neighbors are less than the middle pixel or *z1*. We use the `break` function to stop the routine when this occurs. Essentially, upon each iteration, we use the selected *z1* to find the lowest value pixel from it, which in turn becomes the selected *z1* and so on until the values of the neighborhood are no longer smaller than the selected *z1*.

```
# iterate down the hill until lowest point
while (lowest == FALSE) {
    result <- find_steepest(dem = topo_dem, row_z, col_z) # find
       steepest neighbor
    t <- t + 1
    row_z = result[1]
    col_z = result[2]
    z1 = topo_dem[row_z, col_z]
    transect[t, 1] = row_z
    transect[t, 2] = col_z
    transect[t, 3] = z1
    if (result[3] >= 0)
        {
            lowest == TRUE
            break
        }  # if found lowest point
}
```

Finally we can plot the transect. First lets calculate a distance relative to the top of the transect. After this we can generate a plot as in Fig. 2.16.

```
dist = sqrt((transect[1, 1] - transect[, 1])^2 + (transect[1, 2]
   - transect[, 2])^2)
plot(dist, transect[, 3], type = "l", xlab = "Distance (m)",
   ylab = "Elevation (m)")
```

So let's take this a step further and consider the idea of a random toposequence. In reality, water does not only flow in the steepest direction, water can potentially move down to any lower elevation. And, a toposequence does not necessarily start at the highest elevation either. We can generate a random toposequence (Odgers et al. 2008), where we select a random point in the landscape, then find a random path to

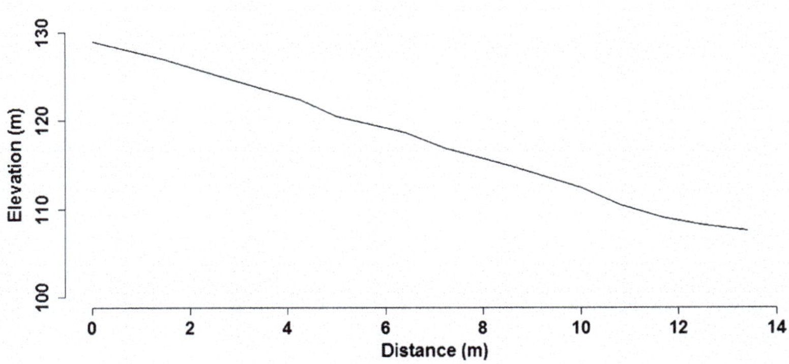

Fig. 2.16 Generated toposequence

the top and bottom of a hillslope. In addition to the downhill routine, we need an uphill routine too.

The algorithm for the random toposequence could be written as:

1. Select a random point from a DEM.
2. Travel uphill:

   2.1 Determine its 3 × 3 neighbor, and determine whether there are higher points?
   2.2 If yes, select randomly a higher point, add to the uphill sequence, and repeat step 2.1. If this point is the highest, the uphill sequence ended.

3. Travel downhill:

   3.1 Determine its 3 × 3 neighbor, and determine whether there are lower points?
   3.2 If yes, select randomly a lower point, add to the downhill sequence, and repeat step 3.1. If this point is the lowest or reached a stream, the downhill sequence ended.

From this algorithm plan, we need to specify two functions, one that allows the transect to travel uphill and another which allows it to travel downhill. For the one to travel downhill, we could use the function from before (`find_steepest`), but we want to build on that function by allowing the user to indicate whether they want a randomly selected smaller value, or whether they want to minimum every time. Subsequently the two new functions would take the following form:

```
# function to simulate water moving down the slope
# input: dem and its row &
travel_down <- function(dem, row_z, col_z, random)
# column random: TRUE use random path, FALSE for steepest path
  return:
# row,col,z-z1 of lower neighbour
{
    z1 = dem[row_z, col_z]
    # find its eight neighbour
    dir = c(-1, 0, 1)
    nr = nrow(dem)
    nc = ncol(dem)
    pz = matrix(data = NA, nrow = 3, ncol = 3)
    for (i in 1:3) {
        for (j in 1:3) {
            ro <- row_z + dir[i]
            co <- col_z + dir[j]
            if (ro > 0 & co > 0 & ro < nr & co < nc) {
                pz[i, j] = dem[ro, co]
            }
        }
    }
    pz[2, 2] = NA
    pz <- pz - z1  # difference with centre value
```

```
    min_pz <- which(pz < 0, arr.ind = TRUE)
    nlow <- nrow(min_pz)
    if (nlow == 0) {
        min_pz <- which(pz == min(pz, na.rm = TRUE),
        arr.ind = TRUE)
    } else {
        if (random) {
            # find random lower value
            ir <- sample.int(nlow, size = 1)
            min_pz <- min_pz[ir, ]
        } else {
            # find lowest value
            min_pz <- which(pz == min(pz, na.rm = TRUE),
            arr.ind = TRUE)
        }
    }
    row_min <- row_z + dir[min_pz[1]]
    col_min <- col_z + dir[min_pz[2]]
    z_min <- dem[row_min, col_min]
    retval <- c(row_min, col_min, min(pz, na.rm = TRUE))
    return(retval)
}

# function to trace water coming from up hill
# input: dem and its row &
travel_up <- function(dem, row_z, col_z, random)
# column random: TRUE use random path, FALSE for steepest path
  return:
# row,col,z-zi of higher neighbour
{
    z1 = dem[row_z, col_z]
    # find its eight neighbour
    dir = c(-1, 0, 1)
    nr = nrow(dem)
    nc = ncol(dem)
    pz = matrix(data = NA, nrow = 3, ncol = 3)
    for (i in 1:3) {
        for (j in 1:3) {
            ro <- row_z + dir[i]
            co <- col_z + dir[j]
            if (ro > 0 & co > 0 & ro < nr & co < nc) {
                pz[i, j] = dem[ro, co]
            }
        }
    }
    pz[2, 2] = NA
    pz <- pz - z1  # difference with centre value

    max_pz <- which(pz > 0, arr.ind = TRUE)  # find higher pixel
    nhi <- nrow(max_pz)
    if (nhi == 0) {
```

```
        max_pz <- which(pz == max(pz, na.rm = TRUE),
        arr.ind = TRUE)
    } else {
        if (random) {
            # find random higher value
            ir <- sample.int(nhi, size = 1)
            max_pz <- max_pz[ir, ]
        } else {
            # find highest value
            max_pz <- which(pz == max(pz, na.rm = TRUE),
            arr.ind = TRUE)
        }
    }
    row_max <- row_z + dir[max_pz[1]]
    col_max <- col_z + dir[max_pz[2]]
    retval <- c(row_max, col_max, max(pz, na.rm = TRUE))
    return(retval)
}
```

Now we can generate a random toposequence. We will use the same `topo_dem` data as before. First we select a point at random using a random selection of a row and column value. Keep in mind that the random point selected here may be different to the one you get because we are using a random number generator via the `sample.int` function.

```
nr <- nrow(topo_dem)  # no. rows in a DEM
nc <- ncol(topo_dem)  # no. cols in a DEM

# start with a random pixel as seed point
row_z1 <- sample.int(nr, 1)
col_z1 <- sample.int(nc, 1)
```

We then can use the `travel_up` function to get our transect to go up the slope.

```
# Travel uphill seed point as a starting point
t <- 1
transect_up <- matrix(data = NA, nrow = 100, ncol = 3)
row_z <- row_z1
col_z <- col_z1
z1 = topo_dem[row_z, col_z]
transect_up[t, 1] = row_z
transect_up[t, 2] = col_z
transect_up[t, 3] = z1

highest = FALSE
# iterate up the hill until highest point
while (highest == FALSE) {
    result <- travel_up(dem = topo_dem, row_z, col_z,
    random = TRUE)
    if (result[3] <= 0)
```

```
        {
            highest == TRUE
            break
        }  # if found lowest point
    t <- t + 1
    row_z = result[1]
    col_z = result[2]
    z1 = topo_dem[row_z, col_z]
    transect_up[t, 1] = row_z
    transect_up[t, 2] = col_z
    transect_up[t, 3] = z1
}
transect_up <- na.omit(transect_up)
```

Next we then use the `travel_down` function to get our transect to go down the slope from the seed point.

```
# travel downhill create a data matrix to store results
transect_down <- matrix(data = NA, nrow = 100, ncol = 3)
# starting point
row_z <- row_z1
col_z <- col_z1
z1 = topo_dem[row_z, col_z]  # a random pixel
t <- 1
transect_down[t, 1] = row_z
transect_down[t, 2] = col_z
transect_down[t, 3] = z1
lowest = FALSE

# iterate down the hill until lowest point
while (lowest == FALSE) {
    result <- travel_down(dem = topo_dem, row_z, col_z,
    random = TRUE)
    if (result[3] >= 0)
        {
            lowest == TRUE
            break
        }  # if found lowest point
    t <- t + 1
    row_z = result[1]
    col_z = result[2]
    z1 = topo_dem[row_z, col_z]
    transect_down[t, 1] = row_z
    transect_down[t, 2] = col_z
    transect_down[t, 3] = z1
}
transect_down <- na.omit(transect_down)
```

The idea then is to bind both uphill and downhill transects into a single one. Note we are using the `rbind` function for this. Furthermore, we are also using the `order` function here to re-arrange the uphill transect so that the resultant binding

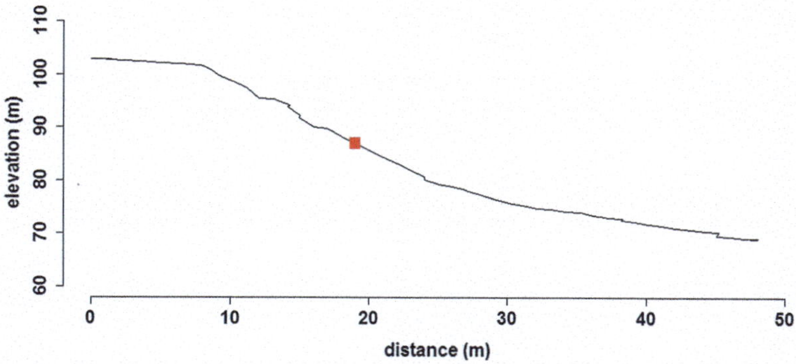

Fig. 2.17 Generated random toposequence. (*Red point* indicates the random seed point)

will be sequential from highest to lowest elevation. Finally, we then calculate the distance relative to the hilltop.

```
transect <- rbind(transect_up[order(transect_up[, 3],
decreasing = T), ],
transect_down[-1, ])
# calculate distance from hilltop
dist = sqrt((transect[1, 1] - transect[, 1])^2 +
(transect[1, 2] - transect[, 2])^2)
```

The last step is to make the plot (Fig. 2.17) of the transect. We can also add the randomly selected seed point for visualization purposes.

```
plot(dist, transect[, 3], type = "l", col = "red",
  xlim = c(0, 100), ylim = c(50, 120), xlab = "Distance (m)",
  ylab = "Elevation (m)")
points(dist[nrow(transect_up)], transect[nrow(transect_up), 3])
```

After seeing how this algorithm works, you can modify the script to take in stream networks, and make the toposequence end once it reaches the stream. You can also add "error trapping" to handle missing values, and also in case where the downhill routine ends up in a local depression. This algorithm also can be used to calculate slope length, distance to a particular landscape feature (e.g. hedges), and so on.

Reference

Odgers NP, McBratney AB, Minasny B (2008) Generation of kth-order random toposequences. Comput Geosci 34(5):479–490

Chapter 3
Getting Spatial in R

R has a very rich capacity to work with, analyse, manipulate and map spatial data. Many procedures one would carry out in a GIS software, can more-or-less be performed relatively easy in R. The application of spatial data analysis in R is well documented in Bivand et al. (2008). Naturally, in DSM, we constantly work with spatial data in one form or another e.g., points, polygons, rasters. We need to do such things as import, view, and export points to, in, and from a GIS. Similarly for polygons and rasters. In this chapter we will cover the fundamentals for doing these basic operations as they are very handy skills, particularly if we want to automate procedures.

Many of the functions used for working with spatial data do not come with the base function suite installed with the R software. Thus we need to use specific functions from a range of different contributed R packages. Probably the most important and most frequently used are:

`sp`	contains many functions for handling vector (polygon) data.
`raster`	very rich source of functions for handling raster data.
`rgdal`	function for projections and spatial data I/O.

Consult the help files and online documentation regarding these packages, and you will quickly realize that we are only scratching the surface of what spatial data analysis functions these few packages are able to perform.

B.P. Malone et al., *Using R for Digital Soil Mapping*,
Progress in Soil Science, DOI 10.1007/978-3-319-44327-0_3

3.1 Basic GIS Operations Using R

3.1.1 Points

We will be working with a small data set of soil information that was collected from the Hunter Valley, NSW in 2010 called HV100. This data set is contained in the ithir package. So first load it in:

```
library(ithir)
data(HV100)
str(HV100)
```

```
## 'data.frame': 100 obs. of  6 variables:
##  $ site: Factor w/ 100 levels "a1","a10","a11",..: 1 ..5..
##  $ x   : num  337860 344060 347035 338235 341760 ...
##  $ y   : num  6372416 6376716 6372741 6368766 6366016 ...
##  $ OC  : num  2.03 2.6 3.42 4.1 3.04 4.07 2.95.. 1.77 ...
##  $ EC  : num  0.129 0.085 0.036 0.081 0.104 0.138 0.07 ...
##  $ pH  : num  6.9 5.1 5.9 6.3 6.1 6.4 5.9 5.5 5.7 6 ...
```

Now load the necessary R packages (you may have to install them onto your computer first):

```
library(sp)
library(raster)
library(rgdal)
```

Using the coordinates function from the sp package we can define which columns in the data frame refer to actual spatial coordinates—here the coordinates are listed in columns x and y.

```
coordinates(HV100) <- ~x + y
str(HV100)
```

```
## Formal class 'SpatialPointsDataFrame' [package "sp"]
   with 5 slots
##    ..@ data        :'data.frame': 100 obs. of  4 variables:
##    .. ..$ site: Factor w/ 100 levels "a1","a10",
    "a11",..: 1 2 3 4 5 ...
##    .. ..$ OC   : num [1:100] 2.03 2.6 3.42 4.1 3.04
    4.07 2.95 3.1 ...
##    .. ..$ EC   : num [1:100] 0.129 0.085 0.036 0.081
    0.104 0.138 0.07....
##    .. ..$ pH   : num [1:100] 6.9 5.1 5.9 6.3 6.1 6.4
    5.9 5.5 5.7 6 ...
##    ..@ coords.nrs : int [1:2] 2 3
##    ..@ coords      : num [1:100, 1:2] 337860 344060 347035..
##    .. ..- attr(*, "dimnames")=List of 2
##    .. .. ..$ : chr [1:100] "1" "2" "3" "4" ...
##    .. .. ..$ : chr [1:2] "x" "y"
```

```
##    ..@ bbox         : num [1:2, 1:2] 335160 6365091 350960
       6382816
##    .. ..- attr(*, "dimnames")=List of 2
##    .. .. ..$ : chr [1:2] "x" "y"
##    .. .. ..$ : chr [1:2] "min" "max"
##    ..@ proj4string:Formal class 'CRS' [package "sp"]
       with 1 slot
##    .. .. ..@ projargs: chr NA
```

Note now that by using the `str` function, the class of `HV100` has now changed from a `dataframe` to a `SpatialPointsDataFrame`. We can do a spatial plot of these points using the `spplot` plotting function in the `sp` package. There are a number of plotting options available, so it will be helpful to consult the help file. Here we are plotting the SOC concentration observed at each location (Fig. 3.1).

```
spplot(HV100, "OC", scales = list(draw = T), cuts = 5,
col.regions = bpy.colors(cutoff.tails = 0.1,
    alpha = 1), cex = 1)
```

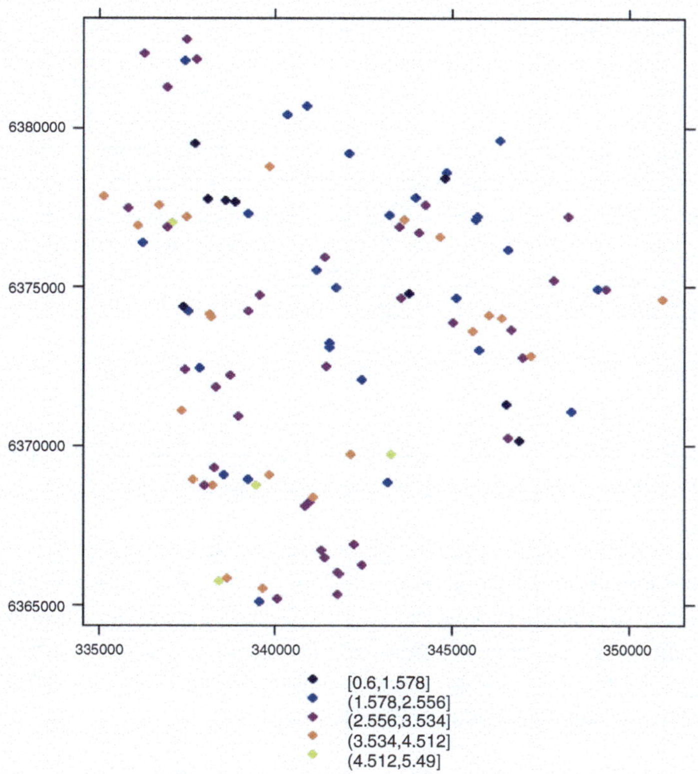

Fig. 3.1 A plot of the site locations with reference to SOC concentration for the 100 points in the `HV100` data set

The SpatialPointsDataFrame structure is essentially the same data frame, except that additional "spatial" elements have been added or partitioned into slots. Some important ones being the bounding box (sort of like the spatial extent of the data), and the coordinate reference system (proj4string), which we need to define for our data set. To define the CRS, we have to know some things about where our data are from, and what was the corresponding CRS used when recording the spatial information in the field. For this data set the CRS used was WGS1984 UTM Zone 56. To explicitly tell R this information we define the CRS as a character string which describes a reference system in a way understood by the PROJ.4 projection library http://trac.osgeo.org/proj/. An interface to the PROJ.4 library is available in the rgdal package. Alternative to using Proj4 character strings, we can use the corresponding yet simpler EPSG code (European Petroleum Survey Group). rgdal also recognizes these codes. If you are unsure of the Proj4 or EPSG code for the spatial data that you have, but know the CRS, you should consult http://spatialreference.org/ for assistance. The EPSG code for texttttWGS1984 UTM Zone 56 is: 32556. So lets define to CRS for this data.

```
proj4string(HV100) <- CRS("+init=epsg:32756")
HV100@proj4string

## CRS arguments:
##  +init=epsg:32756 +proj=utm +zone=56 +south +datum=WGS84
##  +units=m +no_defs +ellps=WGS84 +towgs84=0,0,0
```

We need to define the CRS so that we can perform any sort of spatial analysis. For example, we may wish to use these data in a GIS environment such as Google Earth, ArcGIS, SAGA GIS etc. This means we need to export the SpatialPointsDataFrame of HV100 to an appropriate spatial data format (for vector data) such as a shapefile or KML. rgdal is again used for this via the writeOGR function. To export the data set as a shapefile:

```
writeOGR(HV100, ".", "HV_dat_shape", "ESRI Shapefile")
# Check yor working directory for presence of this file
```

Note that the object we wish to export needs to be a spatial points data frame. You should try opening up this exported shapefile in a GIS software of your choosing.

To look at the locations of the data in Google Earth, we first need to make sure the data is in the WGS84 geographic CRS. If the data is not in this CRS (which is not the case for this data), then we need to perform a coordinate transformation. This is facilitated by using the spTransform function in sp. The EPSG code for WGS84 geographic is: 4326. We can then export out our transformed HV100 data set to a KML file and visualize it in Google Earth.

```
HV100.ll <- spTransform(HV100, CRS("+init=epsg:4326"))
writeOGR(HV100.ll, "HV100.kml", "ID", "KML")
# Check yor working directory for presence of this file
```

Sometimes to conduct further analysis of spatial data, we may just want to import it into R directly. For example, read in a shapefile (this includes both points and polygons too). So lets read in that shapefile that was created just before and saved to the working directory "HV_dat_shape.shp":

```
imp.HV.dat <- readOGR(".", "HV_dat_shape")

## OGR data source with driver: ESRI Shapefile
## Source: ".", layer: "HV_dat_shape"
## with 100 features
## It has 4 fields

imp.HV.dat@proj4string

## CRS arguments:
##  +proj=utm +zone=56 +south +datum=WGS84 +units=m
##  +no_defs +ellps=WGS84 +towgs84=0,0,0
```

The imported shapefile is now a SpatialPointsDataFrame, just like the HV100 data that was worked on before, and is ready for further analysis.

3.1.2 *Rasters*

Most of the functions needed for handling raster data are contained in the raster package. There are functions for reading and writing raster files from and to different raster formats. In DSM we work quite a deal with data in table format and then rasterise this data so that we can make a map. To do this in R, lets bring in a data frame. This could be either from a text-file, but as for the previous occasions the data is imported from the ithir package. This data is a digital elevation model with 100 m grid resolution, from the Hunter Valley, NSW, Australia.

```
library(ithir)
data(HV_dem)
str(HV_dem)

## 'data.frame': 21518 obs. of  3 variables:
##  $ X        : num  340210 340310 340110 340210...
##  $ Y        : num  6362641 6362641 6362741 6362741...
##  $ elevation: num  177 175 178 172 173 ...
```

As the data is already a raster (such that the row observation indicate locations on a regular spaced grid), but in a table format, we can just use the rasterFromXYZ function from raster. Also we can define the CRS just like we did with the HV100 point data we worked with before.

```
r.DEM <- rasterFromXYZ(HV_dem)
proj4string(r.DEM) <- CRS("+init=epsg:32756")
```

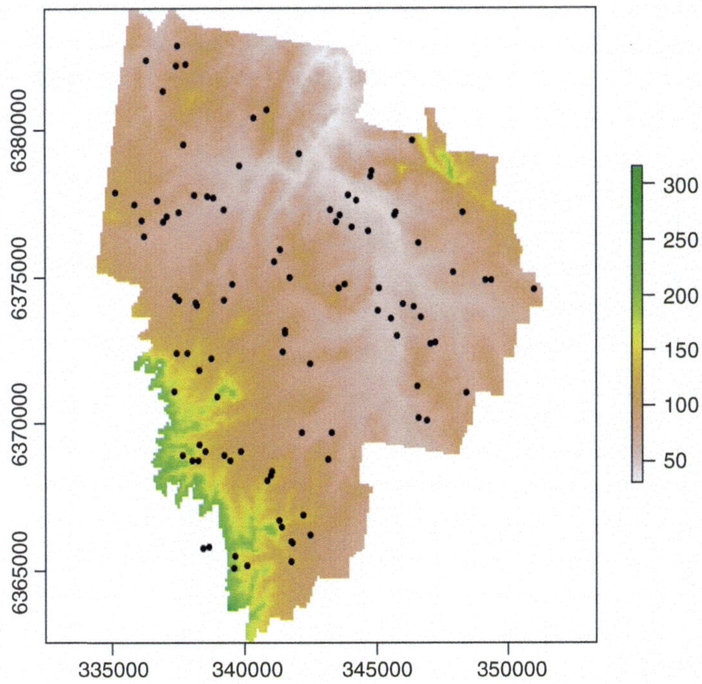

Fig. 3.2 Digital elevation model for the Hunter Valley, overlayed with the HV100 sampling sites

So lets do a quick plot of this raster and overlay the HV100 point locations (Fig. 3.2).

```
plot(r.DEM)
points(HV100, pch = 20)
```

So we may want to export this raster to a suitable format for further work in a standard GIS environment. See the help file for `writeRaster` to get information regarding the supported grid types that data can be exported. For demonstration, we will export our data to ESRI Ascii ascii, as it is a common and universal raster format.

```
writeRaster(r.DEM, filename = "HV_dem100.asc",
format = "ascii", overwrite = TRUE)
```

What about exporting raster data to KML file? Here you could use the KML function. Remember that we need to reproject our data because it is in the UTM system, and need to get it to WGS84 geographic. The raster re-projection is performed using the `projectRaster` function. Look at the help file for this function. Probably the most important parameters are `crs`, which takes the CRS string of the projection you want to convert the existing raster to, assuming it already has a defined CRS. The other is `method` which controls the interpolation method. For continuous data, "bilinear" would be suitable, but for categorical, "ngb", (which is nearest neighbor interpolation) is probably better suited. `KML` is a handy function from `raster` for exporting grids to kml format.

```
p.r.DEM <- projectRaster(r.DEM, crs = "+init=epsg:4326",
method = "bilinear")

KML(p.r.DEM, "HV_DEM.kml", col = rev(terrain.colors(255)),
overwrite = TRUE)
# Check yor working directory for presence of the kml file
```

Now visualize this in Google Earth and overlay this map with the points that were created before.

The other useful procedure we can perform is to import rasters directly into `R` so we can perform further analyses. `rgdal` interfaces with the GDAL library, which means that there are many supported grid formats that can be read into `R` http://www.gdal.org/formats_list.html. Here we will load in the "HV_dem100.asc" raster that was made just before.

```
read.grid <- readGDAL("HV_dem100.asc")

## HV_dem100.asc has GDAL driver AAIGrid
## and has 215 rows and 169 columns
```

The imported raster `read.grid` is a `SpatialGridDataFrame`, which is a formal class of the `sp` package. To be able to use the raster functions from `raster` we need to convert it to the `RasterLayer` class.

```
grid.dem <- raster(read.grid)
grid.dem

## class       : RasterLayer
## dimensions  : 215, 169, 36335  (nrow, ncol, ncell)
## resolution  : 100, 100  (x, y)
## extent      : 334459.8, 351359.8, 6362591, 6384091
                 (xmin, xmax, ymin, ymax)
## coord. ref. : NA
## data source : in memory
## names       : band1
## values      : 29.61407, 315.6837  (min, max)
```

You will notice from the `R` generated output indicating the data source, it says it is loaded into memory. This is fine for small rasters, but can become a problem when very large rasters need to be handled. A really powerful feature of the `raster` package is the ability to point to the location of a raster/s without the need to load it into memory. It is only very rarely that one needs to use all the data contained in a raster at one time. As will be seen later on this useful feature makes for a very efficient way to perform digital soil mapping across very large spatial extents.

To point to the "HV_dem100.asc" raster that was created earlier we would use the following or similar command (where `getwd()` is the function to return the address string of the working directory):

```
grid.dem <- raster(paste(paste(getwd(), "/", sep = ""),
"HV_dem100.asc", sep = ""))
grid.dem
```

```
## class       : RasterLayer
## dimensions  : 215, 169, 36335  (nrow, ncol, ncell)
## resolution  : 100, 100  (x, y)
## extent      : 334459.8, 351359.8, 6362591, 6384091
                 (xmin, xmax, ymin, ymax)
## coord. ref. : NA
## data source : C:\Users\bmalone\Dropbox\2015\DSM_book\
                 HV_dem100.asc
## names       : HV_dem100
```

```
# plot(grid.dem)
```

3.2 Advanced Work: Creating Interactive Maps in R

A step beyond creating kml files of your digital soil information is the creation of customized interactive mapping products that can be visualized within your web browser. Interactive mapping makes sharing your data with colleagues simpler, and importantly improves the visualization experience via customization features that are difficult to achieve via the Google Earth software platform. The interactive mapping is made possible via the Leaflet R package. Leaflet is one of the most popular open-source JavaScript libraries for interactive maps. The Leaflet R package makes it easy to integrate and control Leaflet maps in R. More detailed information about Leaflet can be found at http://leafletjs.com/, and information specifically about the R package is at https://rstudio.github.io/leaflet/.

There is a common workflow for creating Leaflet maps in R. First is the creation of a map widget (calling `leaflet()`); followed by the adding of layers or features to the map by using layer functions (e.g. `addTiles`, `addMarkers`, `addPolygons`) to modify the map widget. The map can then be printed and visualized in the R image window or saved to HTML file for visualization within a web browser. The following R script is a quick taste of creating an interactive Leaflet map. It is assumed that the `leaflet` and `magrittr` are installed.

```
library(leaflet)
library(magrittr)
```

```
leaflet() %>% addMarkers(lng = 151.210558, lat = -33.852543,
```

```
popup = "The view from here is amazing!") %>%
addProviderTiles("Esri.WorldImagery")
```

You should now see in your plot window a map of an iconic Australian landmark. Interactive features of this map include markers with text, plus ability to zoom and map panning. More will be discussed about the layer functions of the leaflet map further on. What has not been encountered yet is the forward pipe operator %>%. This operator will forward a value, or the result of an expression, into the next function call or expression. To use this operator the `magrittr` package is required. The example script below shows the same example using and not using the forward pipe operator.

```
# Draw 100 random uniformly distributed numbers between 0 and 1
x <- runif(100)

sqrt(sum(range(x)))

## ..is equivalent to (using forward pipe operator)
x %>% range %>% sum %>% sqrt
```

Sometimes what we want to do in R can get lost within a jumble of brackets, whereas using the forward pipe operator the process of operations is a lot clearer. So lets begin to construct some Leaflet mapping using the data from a little earlier regarding the point (`HV100.11`) and raster data (`p.r.DEM`). Note that these data have already been converted to WGS coordinate reference system, which is necessary for the creation of the interactive mapping outputs. Firstly, lets create a basic map—example of not using and then using the forward pipe operator.

```
# Basic map without piping operator
addMarkers(addTiles(leaflet()), data = HV100.11)

# with forward pipe operator
leaflet() %>% addTiles() %>% addMarkers(data = HV100.11)
```

With the above, we are calling upon a pre-existing base map via the `addTiles()` function. Leaflet supports base maps using map tiles, popularized by Google Maps and now used by nearly all interactive web maps. By default, OpenStreetMap https://www.openstreetmap.org/#map=13/-33.7760/150.6528&layers=C tiles are used. Alternatively, many popular free third-party base maps can be added using the `addProviderTiles()` function, which is implemented using the leaflet-providers plugin. For example, previously we used the `Esri.WorldImagery` base mapping. The full set of possible base maps can be found at http://leaflet-extras.github.io/leaflet-providers/preview/index.html. Note that an internet connection is required for access to the base maps and map tiling. The last function used above the `addMarkers` function, we simply call up the point data we used previously, which are those soil point observations and measurements from the Hunter Valley, NSW. A basic map will have been created with your plot window. For the next step, lets populate the markers we have created with some of the data that was measured, then add the `Esri.WorldImagery` base mapping.

```
# Populate pop-ups
my_pops <- paste0("<strong>Site: </strong>", HV100.ll$site,
"<br>\n  <strong> Organic Carbon (%): </strong>",
HV100.ll$OC, "<br>\n  <strong> soil pH: </strong>", HV100.ll$pH)

# Create interactive map
leaflet() %>% addProviderTiles("Esri.WorldImagery") %>%
addMarkers(data = HV100.ll, popup = my_pops)
```

Further, we can colour the markers and add a map legend. Here we will get the quantiles of the measured SOC% and color the markers accordingly. Note that you will need the colour ramp package `RColorBrewer` installed.

```
library(RColorBrewer)

# Colour ramp
pal1 <- colorQuantile("YlOrBr", domain = HV100.ll$OC)

# Create interactive map
leaflet() %>%
addProviderTiles("Esri.WorldImagery") %>%
addCircleMarkers(data = HV100.ll, color = ~pal1(OC),
popup = my_pops) %>%
addLegend("bottomright", pal = pal1,
values = HV100.ll$OC, title = "Soil Organic Carbon (%) quantiles",
opacity = 0.8)
```

It is very worth consulting the help files associated with the `leaflet` R package for further tips on creating further customized maps. The website dedicated to that package, which was mentioned above is also a very helpful resource too.

Raster maps can also be featured in our interactive mapping too, as illustrated in the following script.

```
# Colour ramp
pal2 <- colorNumeric(brewer.pal(n = 9, name = "YlOrBr"),
domain = values(p.r.DEM), na.color = "transparent")

# interactive map
leaflet() %>%
addProviderTiles("Esri.WorldImagery") %>%
addRasterImage(p.r.DEM, colors = pal2, opacity = 0.7) %>%
addLegend("topright", opacity = 0.8, pal = pal2,
values = values(p.r.DEM), title = "Elevation")
```

Lastly, we can create an interactive map that allows us to switch between the different mapping outputs that we have created.

```
# layer switching
leaflet() %>%
addTiles(group = "OSM (default)") %>%
addProviderTiles("Esri.WorldImagery") %>%

addCircleMarkers(data = HV100.ll, color = ~pal1(OC),
group = "points", popup = my_pops) %>%
addRasterImage(p.r.DEM, colors = pal2, group = "raster",
opacity = 0.8) %>%
addLayersControl(baseGroups = c("OSM (default)", "Imagery"),
overlayGroups = c("points", "raster"))
```

With the created interactive mapping, we can then export these as a web page in HTML format. This can be done via the export menu within the R-Studio plot window, where you want to select the option for "Save as Web page". This file can then be easily shared and viewed by your colleagues.

3.3 Some R Packages That Are Useful for Digital Soil Mapping

Notwithstanding to the rich statistical and analytical resource provided through the R base functionality, the following R packages (and their contained functions) are what we think are an invaluable resource for DSM. As with all things in R, one discovers new tricks all the time, which subsequently means that what functions and analyses are useful now, are superseded or made obsolete later on. There are four main groups of tasks that are critical for implementing DSM in general. These are: (1) Soil science and pedometric type tasks; (2) Using GIS tools and related GIS tasks; (4) Modelling; (4) Making maps, plotting etc. The following are short introductions about those packages that fall into these categories.

Soil science and pedometrics

- `aqp`: Algorithms for quantitative pedology. http://cran.r-project.org/web/packages/aqp/index.html. A collection of algorithms related to modeling of soil resources, soil classification, soil profile aggregation, and visualization.
- `GSIF`: Global soil information facility. http://cran.r-project.org/web/packages/GSIF/index.html. Tools, functions and sample datasets for digital soil mapping.

GIS

- `sp`: http://cran.r-project.org/web/packages/sp/index.html. A package that provides classes and methods for spatial data. The classes document where the spatial location information resides, for 2D or 3D data. Utility functions are provided, e.g. for plotting data as maps, spatial selection, as well as methods for retrieving coordinates, for sub-setting, print, summary, etc.

- `raster`: http://cran.r-project.org/web/packages/raster/index.html. Reading, writing, manipulating, analyzing and modeling of gridded spatial data. The package implements basic and high-level functions and processing of very large files is supported.
- `rgdal`: http://cran.r-project.org/web/packages/rgdal/index.html. Provides bindings to Frank Warmerdam's Geospatial Data Abstraction Library (GDAL) (>= 1.6.3) and access to projection/transformation operations from the PROJ.4 library. Both GDAL raster and OGR vector map data can be imported into `R`, and GDAL raster data and OGR vector data exported. Use is made of classes defined in the `sp` package.
- `RSAGA`: http://cran.r-project.org/web/packages/RSAGA/index.html. RSAGA provides access to geocomputing and terrain analysis functions of SAGA GIS http://www.saga-gis.org/en/index.html from within `R` by running the command line version of SAGA. RSAGA furthermore provides several `R` functions for handling ASCII grids, including a flexible framework for applying local functions (including predict methods of fitted models) and focal functions to multiple grids.

Modeling

- `caret`: http://cran.r-project.org/web/packages/caret/index.html. Extensive range of functions for training and plotting classification and regression models. See the caret website for more detailed information http://topepo.github.io/caret/index.html.
- `Cubist`: http://cran.r-project.org/web/packages/Cubist/index.html. Regression modeling using rules with added instance-based corrections. Cubist models were developed by Ross Quinlan. Further information can be found at Rulequest https://www.rulequest.com/
- `C5.0`: http://cran.r-project.org/web/packages/C50/index.html. C5.0 decision trees and rule-based models for pattern recognition. Another model structure developed by Ross Quinlan.
- `gam`: http://cran.r-project.org/web/packages/gam/index.html. Functions for fitting and working with generalized additive models.
- `nnet`: http://cran.r-project.org/web/packages/nnet/index.html. Software for feed-forward neural networks with a single hidden layer, and for multinomial log-linear models.
- `gstat`: http://cran.r-project.org/web/packages/gstat/. Variogram modelling; simple, ordinary and universal point or block (co)kriging, sequential Gaussian or indicator (co)simulation; variogram and variogram map plotting utility functions. A related and useful package is `automap` (http://cran.r-project.org/web/packages/automap/index.html), which performs an automatic interpolation by automatically estimating the variogram and then calling `gstat`.

Mapping and plotting

- Both `raster` and `sp` have handy functions for plotting spatial data. Besides using the base plotting functionality, another useful plotting package is `ggplot2` (http://cran.r-project.org/web/packages/ggplot2/index.html). This package is an implementation of the grammar of graphics in R. It combines the advantages of both base and lattice graphics: conditioning and shared axes are handled automatically, and you can still build up a plot step by step from multiple data sources. It also implements a sophisticated multidimensional conditioning system and a consistent interface to map data to aesthetic attributes. See the ggplot2 website for more information, documentation and examples (http://ggplot2.org/).

Reference

Bivand RS, Pebesma EJ, Gomez-Rubio V (2008) Applied spatial data analysis with R. UseR! series. Springer, New York

Chapter 4
Preparatory and Exploratory Data Analysis for Digital Soil Mapping

At the start of this book, some of the history and theoretical underpinnings of DSM was discussed. Now with a solid foundation in R, it is time to put this all into practice i.e. do DSM with R.

In this chapter some common methods for soil data preparation and exploration are covered. Soil point databases are inherently heterogeneous because soils are measured non uniformly from site to site. However one more-or-less commonality is that soil point observations will generally have some sort of label, together with some spatial coordinate information that indicates where the sample was collected from. Then things begin to vary from site to site. Probably the biggest difference is that all soils are not measured universally at the same depths. Some soils are sampled per horizon or at regular depths. Some soil studies examine only the topsoil, while others sample to the bedrock depth. Then different soil attributes are measured at some locations and depths, but not at others. Overall, it becomes quickly apparent when one begins working with soil data that a number of preprocessing steps are needed to fulfill the requirements of a particular analysis.

In order to prepare a collection of data for use in a DSM project as described in Minasny and McBratney (2010) one needs to examine what data are available, what is the soil attribute or class to be modeled? What is the support of the data? This includes whether observations represent soil point observations or some integral over a defined area (for now we just consider observations to be point observations). However, we may also assume the support to be also a function of depth in that we may be interested in only mapping soil for the top 10 cm, or to 1 m, or any depth in between or to the depth to bedrock. The depth interval could be a single value (such as one value for the 0–1 m depth interval as an example), or we may wish to map simultaneously the depth variation of the target soil attribute with the lateral or spatial variation. These questions add complexity to the soil mapping project, but are an important consideration when planning a project and assessing what the objectives are.

B.P. Malone et al., *Using R for Digital Soil Mapping*,
Progress in Soil Science, DOI 10.1007/978-3-319-44327-0_4

More recent digital soil mapping research has examined the combination of soil depths functions with spatial mapping in order to create soil maps with a near 3-D support. In the following section some approaches for doing this are discussed with emphasis and instruction on a particular method, namely the use of a mass-preserving soil depth function. This will be followed by a section that will examine the important DSM step of linking observed soil information with available environmental covariates and the subsequent preparation of covariates for spatial modelling.

4.1 Soil Depth Functions

The traditional method of sampling soil involves dividing a soil profile into horizons. The number of horizons and the position of each are generally based on attributes easily observed in the field, such as morphological soil properties (Bishop et al. 1999). From each horizon, a bulk sample is taken and it is assumed to represent the average value for a soil attribute over the depth interval from which it is sampled. There are some issues with this approach, particularly from a pedological perspective and secondly from the difficulty in using this legacy data within a Digital Soil Mapping (DSM) framework where we wish to know the continuous variability of a soil both in the lateral and vertical dimensions. From the pedological perspective soil generally varies continuously with depth; however, representing the soil attribute value as the average over the depth interval of horizons leads to discontinuous or stepped profile representations. Difficulties can arise in situations where one wants to know the value of an attribute at a specified depth. The second issue is regarding DSM and is where we use a database of soil profiles to generate a model of soil variability in the area in which they exist. Because observations at each horizon for each profile will rarely be the same between any two profiles, it then becomes difficult to build a model where predictions are made at a set depth or at standardized depth intervals.

Legacy soil data is too valuable to do away with and thus needs to be molded to suit the purposes of the map producer, such that one needs to be able to derive a continuous function using the available horizon data as some input. This can be done with many methods including polynomials and exponential decay type depth functions. A more general continuous depth function is the equal-area quadratic spline function. The usage and mathematical expression of this function have been detailed in Ponce-Hernandez et al. (1986), Bishop et al. (1999), and Malone et al. (2009). A useful feature of the spline function is that it is mass preserving, or in other words the original data is preserved and can be retrieved again via integration of the continuous spline. Compared to exponential decay functions where the goal is in defining the actual parameters of the decay function, the spline parameters are the values of the soil attribute at the standard depths that are specified by the user. This is a useful feature, because firstly, one can harmonize a whole collection of soil profile data and then explicitly model the soil for a specified depth. For example the GlobalSoilMap.net project (Arrouays et al. 2014) has a specification that digital soil maps be created for each target soil variable for the 0–5, 5–15, 15–30, 30–60, 60–100, and 100–200 cm depth intervals. In this case, the mass-preserving splines can be fitted to the observed data, then values can be extracted from them at the required depths, and are then ready for exploratory analysis and spatial modelling.

In the following, we will use legacy soil data and the spline function to prepare data to be used in a DSM framework. This will specifically entail fitting splines to all the available soil profiles and then through a process of harmonization, integrate the splines to generate harmonized depths of each observation.

4.1.1 Fit Mass Preserving Splines with R

We will demonstrate the mass-preserving spline fitting using a single soil profile example for which there are measurements of soil carbon density to a given maximum depth. We can fit a spline to the maximum soil depth, or alternatively any depth that does not exceed the maximum soil depth. The function used for fitting splines is called `ea_spline` and is from the `ithir` package. Look at the help file for further information on this function. For example, there is further information about how the `ea_spline` function can also accept data of class `SoilProfileCollection` from the aqp package in addition to data of the more generic `data.frame` class. In the example below the input data is of class `data.frame`. The data we need (`oneProfile`) is in the `ithir` package.

```
library(ithir)
data(oneProfile)
str(oneProfile)
```

```
## 'data.frame': 8 obs. of  4 variables:
##  $ Soil.ID        : num  1 1 1 1 1 1 1 1
##  $ Upper.Boundary: num  0 10 30 50 70 120 250 350
##  $ Lower.Boundary: num  10 20 40 60 80 130 260 360
##  $ C.kg.m3.       : num  20.7 11.7 8.2 6.3 2.4 2 0.7
```

As you can see above, the data table shows the soil depth information and carbon density values for a single soil profile. Note the discontinuity of the observed depth intervals which can also be observed in Fig. 4.1.

The `ea_spline` function will predict a continuous function from the top of the soil profile to the maximum soil depth, such that it will interpolate values both within the observed depths and between the depths where there is no observation.

To parametize the `ea_spline` function, we could accept the defaults, however it might be likely to change the `lam` and `d` parameters to suit the objective of the analysis being undertaken. `lam` is the lambda parameter which controls the

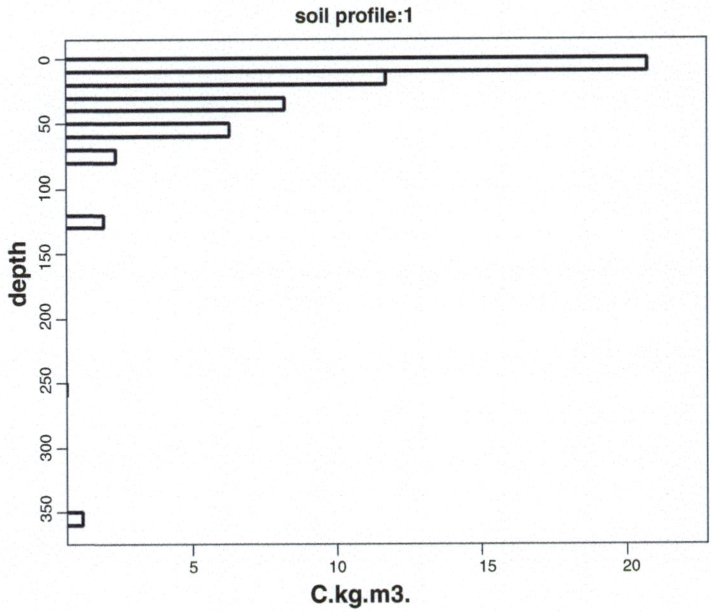

Fig. 4.1 Soil profile plot of the `oneProfile` data. Note this figure was produced using the `plot_soilProfile` function from `ithir`

smoothness or fidelity of the spline. Increasing this value will make the spline more rigid. Decreasing it towards zero will make the spline more flexible such that it will follow near directly the observed data. A sensitivity analysis is generally recommended in order to optimize this parameter. From experience a `lam` value of 0.1 works well generally for most soil properties, and is the default value for the function. The `d` parameter represents the depth intervals at which we want to get soil values for. This is a harmonization process where regardless of which depths soil was observed at, we can derive the soil values for regularized and standard depths. In practice, the continuous spline function is first fitted to the data, then we get the integrals of this function to determine the values of the soil at the standard depths. `d` is a matrix, but on the basis of the default values, what it is indicating is that we want the values of soil at the following depth intervals: 0–5, 5–15, 15–30, 30–60, 60–100, and 100–200 cm. These depths are specified depths determined for the GlobalSoilMap.net project (Arrouays et al. 2014). Naturally, one can alter these values to suit there own particular requirements. To fit a spline to the carbon density values of the `oneProfile` data, the following script could be used:

```
eaFit <- ea_spline(oneProfile, var.name = "C.kg.m3.",
    d = t(c(0, 5, 15, 30, 60, 100, 200)), lam = 0.1, vlow = 0,
    show.progress = FALSE)
```

```
## Fitting mass preserving splines per profile...
str(eaFit)
```

```
## List of 4
##  $ harmonised:'data.frame': 1 obs. of  8 variables:
##   ..$ id         : num 1
##   ..$ 0-5 cm     : num 21
##   ..$ 5-15 cm    : num 15.8
##   ..$ 15-30 cm   : num 9.89
##   ..$ 30-60 cm   : num 7.18
##   ..$ 60-100 cm  : num 2.76
##   ..$ 100-200 cm: num 1.73
##   ..$ soil depth: num 360
##  $ obs.preds :'data.frame': 8 obs. of  6 variables:
##   ..$ Soil.ID       : num [1:8] 1 1 1 1 1 1 1 1
##   ..$ Upper.Boundary: num [1:8] 0 10 30 50 70 120 250 350
##   ..$ Lower.Boundary: num [1:8] 10 20 40 60 80 130 260 360
##   ..$ C.kg.m3.      : num [1:8] 20.7 11.7 8.2 6.3 2.4
                        2 0.7 1.2
##   ..$ predicted     : num [1:8] 19.84 12.45 8.24 6.2 2.56 ...
##   ..$ FID           : num [1:8] 1 1 1 1 1 1 1 1
##  $ var.1cm   : num [1:200, 1] 21.6 21.4 21.1 20.8 20.3 ...
##  $ tmse      : num [1, 1] 0.263
```

The output of the function is a `list`, where the first element is a `dataframe` (`harmonized`) which are the predicted spline estimates at the specified depth intervals. The second element (`obs.preds`) is another `dataframe` but contains the observed soil data together with spline predictions for the actual depths of observation for each soil profile. The third element (`var.1cm`) is a matrix which stores the spline predictions of the depth function at (in this case) 1 cm resolution. Each column represent a given soil profile and each row represents an incremental 1 cm depth increment to the maximum depth we wish to extract values for, or to the maximum observed soil depth (which ever is smallest). The last element (`tmse`) is another matrix but stores a single mean square error estimate for each given soil profile. This value is an estimate of the magnitude of difference between observed values and associated predicted values with each profile. It is often more amenable to visualize the performance of the spline fitting. Subsequently, plotting the outputs of `ea_spline` is made possible by the associated `plot_ea_spline` function (see help file for use of this function):

```
par(mfrow = c(3, 1))
for (i in 1:3) {
    plot_ea_spline(splineOuts = eaFit, d = t(c(0, 5, 15, 30, 60,
    100, 200)), maxd = 200, type = i, plot.which = 1,
    label = "carbon density") }
```

The `plot_ea_spline` function is a basic function without too much control over the plotting parameters, except there are three possible themes of plot output that one can select. This is controlled by the `type` parameter. `Type = 1` is to return the observed soil data plus the continuous spline (default). `Type = 2` is to return the observed data plus the averages of the spline at the specified depth intervals. `Type = 3` is to return the observed data, spline averages and continuous spline. The script above results in producing all three possible plots, and is shown on Fig. 4.2.

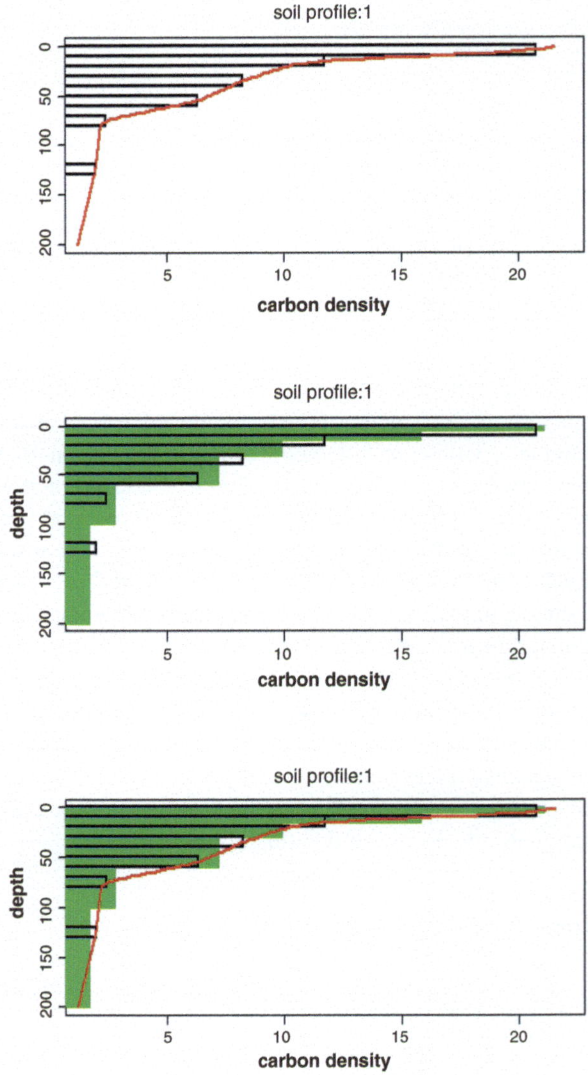

Fig. 4.2 Soil profile plot of the three type variants from `plot_ea_spline`. Plot 1 is type 1, plot 2 is type 2 and plot 3 is type 3

4.2 Intersecting Soil Point Observations with Environmental Covariates

In order to carry out digital soil mapping in terms of evaluating the significance of environmental variables in explaining the spatial variation of the target soil variable under investigation, we need to link both sets of data together and extract the values of the covariates at the locations of the soil point data. The first task is to bring in to our working environment some soil point data. We will be using a preprocessed data set of the Edgeroi Data set (McGarry et al. 1989) with the target variable being soil carbon density. The data was preprocessed such that the predicted values are outputs of the mass-preserving depth function. The data is loaded in from the `ithir` package with the following script:

```
data(edgeroi_splineCarbon)
```

```
## 'data.frame': 341 obs. of  10 variables:
##  $ id          : Factor w/ 341 levels "ed001","ed002",..:
                   1 2 3 4
##  $ east        : num  741912 744662 747412 750212 752912 ...
##  $ north       : num  6678083 6677983 6677933 6677933
                   6677883 ...
##  $ X0.5.cm     : num  28.8 23.7 12.5 7.7 7.5 8.7 17.3 11.7 ...
##  $ X5.15.cm    : num  18.6 14.7 9.3 6.3 6.9 7.8 12.6 ...
##  $ X15.30.cm   : num  10.5 6.5 6.6 5.5 6.5 6.6 8 8.7 5.7 8 ...
##  $ X30.60.cm   : num  9.2 6.6 5.6 5.6 6 5.9 7.3 6.6 5.7 7.2 ...
##  $ X60.100.cm  : num  5.3 6.1 4 4.1 4.4 4.8 3.9 3.3 4.5 4.4 ...
##  $ X100.200.cm: num  3.1 3.3 1.2 1.8 2.7 1.9 1.3 2.4
                   1.7 0.8 ...
##  $ soil.depth  : num  260 260 260 260 260 259 260 260
                   253 260 ...
```

As the summary shows above in the column headers, the soil depths correspond to harmonized depth intervals. Before we create a spatial plot of these data, it is a good time to introduce some environmental covariates. A small subset of them is available for the whole Edgeroi District at a consistent pixel resolution of 90 m. These can be accessed using the script:

```
data(edgeroiCovariates)
```

This data set stores 5 `rasterLayers` which correspond to: `elevation`, terrain wetness index (`twi`), gamma radiometric potassium (`radK`), and Landsat 7 spectral bands 3 and 4 (`landsat_b3` and `landsat_b4`). For example, we can call up the metadata that corresponds to each of these rasters using:

```
library(raster)
elevation
```

```
## class        : RasterLayer
## dimensions   : 400, 577, 230800  (nrow, ncol, ncell)
## resolution   : 90, 90  (x, y)
## extent       : 738698.6, 790628.6, 6643808, 6679808
                  (xmin, xmax, ymin, ymax)
## coord. ref. : +proj=utm +zone=55 +south +ellps=WGS84
## +datum=WGS84 +units=m +no_defs
## data source : in memory
## names        : elevation
## values       : 181.4204, 960.1074  (min, max)
```

twi

```
## class        : RasterLayer
## dimensions   : 400, 577, 230800  (nrow, ncol, ncell)
## resolution   : 90, 90  (x, y)
## extent       : 738698.6, 790628.6, 6643808, 6679808
                  (xmin, xmax, ymin, ymax)
## coord. ref. : +proj=utm +zone=55 +south +ellps=WGS84
## +datum=WGS84 +units=m +no_defs
## data source : in memory
## names        : twi
## values       : 9.801188, 23.89634  (min, max)
```

radK

```
## class        : RasterLayer
## dimensions   : 400, 577, 230800  (nrow, ncol, ncell)
## resolution   : 90, 90  (x, y)
## extent       : 738698.6, 790628.6, 6643808, 6679808
                  (xmin, xmax, ymin, ymax)
## coord. ref. : +proj=utm +zone=55 +south +ellps=WGS84
## +datum=WGS84 +units=m +no_defs
## data source : in memory
## names        : radK
## values       : -0.00929, 5.16667  (min, max)
```

landsat_b3

```
## class        : RasterLayer
## dimensions   : 400, 577, 230800  (nrow, ncol, ncell)
## resolution   : 90, 90  (x, y)
## extent       : 738698.6, 790628.6, 6643808, 6679808
                  (xmin, xmax, ymin, ymax)
## coord. ref. : +proj=utm +zone=55 +south +ellps=WGS84
## +datum=WGS84 +units=m +no_defs
## data source : in memory
## names        : landsat_b3
## values       : 18.86447, 170.517  (min, max)
```

```
landsat_b4

## class        : RasterLayer
## dimensions   : 400, 577, 230800  (nrow, ncol, ncell)
## resolution   : 90, 90  (x, y)
## extent       : 738698.6, 790628.6, 6643808, 6679808
                  (xmin, xmax, ymin, ymax)
## coord. ref. : +proj=utm +zone=55 +south +ellps=WGS84
## +datum=WGS84 +units=m +no_defs
## data source : in memory
## names        : landsat_b4
## values       : 13.18422, 154.5758  (min, max)
```

You may notice that there is a commonality between these `rasterlayers` in terms of their CRS, dimensions and resolution. This harmony is an ideal situation for DSM where there may often be instances where rasters from the some area under investigation may have different resolutions, extents and even CRSs. It these situations it is common to reproject and or resample to a common projection and resolution. The functions from the `raster` package which may be of use in these situations are `projectRaster` and `resample`.

While the example is a little contrived, it is useful to always determine whether or not the available covariates have complete coverage of the soil point data. This might be done with the following script which will produce a figure like in Fig. 4.3:

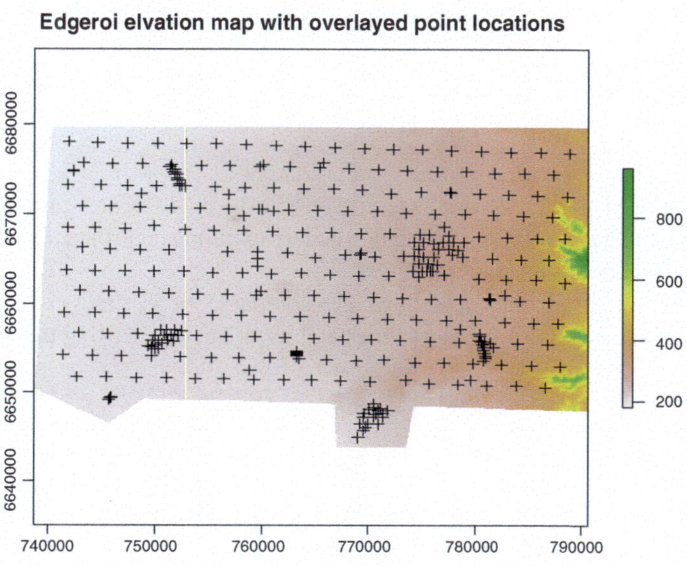

Fig. 4.3 Edgeroi elevation map with the soil point locations overlayed upon it

```
# plot raster
plot(elevation, main = "Edgeroi elvation map with overlayed
     point locations")


coordinates(edgeroi_splineCarbon) <- ~east + north


## plot points
plot(edgeroi_splineCarbon, add = T)
```

When the covariate data is of common resolution and extent, rather than working with each raster independently it is much more efficient to stack them all into a single object. The `stack` function from `raster` is ready-made for this, and is simple enacted with the following script:

```
covStack <- stack(elevation, twi, radK, landsat_b3, landsat_b4)
covStack

## class       : RasterStack
## dimensions  : 400, 577, 230800, 5  (nrow, ncol, ncell, nlayers)
## resolution  : 90, 90  (x, y)
## extent      : 738698.6, 790628.6, 6643808, 6679808
                 (xmin, xmax, ymin, ymax)
## coord. ref. : +proj=utm +zone=55 +south +ellps=WGS84
## +datum=WGS84 +units=m +no_defs
## names       : elevation,         twi,        radK,
                 landsat_b3, landsat_b4
## min values  : 181.420395,    9.801188,   -0.009290,
                 18.864470,  13.184220
## max values  : 960.10742,    23.89634,     5.16667,
                 170.51700,  154.57581
```

As mentioned earlier, it is always preferable to have all the rasters you are working with to have a common spatial extent, resolution and projection. Otherwise the `stack` function will encounter an error.

With the soil point data and covariates prepared, it is time to perform the intersection between the soil observations and covariate layers using the script:

```
DSM_data <- extract(covStack, edgeroi_splineCarbon, sp = 1,
method = "simple")
```

The `extract` function is quite useful. Essentially the function ingests the `rasterStack` object, together with the `SpatialPointsDataFrame` object `edgeroi_splineCarbon`. The `sp` parameter set to 1 means that the extracted covariate data gets appended to the existing `SpatialPointsDataFrame` object. While the `method` object specifies the extraction method which in our case is "simple" which likened to get the covariate value nearest to the points i.e it is likened to "drilling down".

A good practice is to then export the soil and covariate data intersect object to file for later use. First we convert the spatial object to a `dataframe`, then export as a comma separated text file.

```
DSM_data <- as.data.frame(DSM_data)
write.table(DSM_data, "edgeroiSoilCovariates_C.TXT",
col.names = T, row.names = FALSE, sep = ",")
```

4.2.1 *Using Rasters from File*

In the previous example the rasters we wanted to use are available data from the `ithir` package. More generally we will have the raster data we need sitting on our computer or disk somewhere. The steps for intersecting the soil observation data with the covariates are the same as before, except we now need to specify the location where our raster covariate data is located. We need not even have to load in the rasters to memory, just point `R` to where they are, and then run the raster `extract` function. This utility is obviously a very handy feature when we are dealing with an inordinately large or large number of rasters. The work function we need is `list.files`. For example:

```
list.files(path = "C:/temp/testGrids/", pattern = "\\.tif$",
full.names = TRUE)

## [1] "C:/temp/testGrids/edge_elevation.tif"
## [2] "C:/temp/testGrids/edge_landsat_b3.tif"
## [3] "C:/temp/testGrids/edge_landsat_b4.tif"
## [4] "C:/temp/testGrids/edge_radK.tif"
## [5] "C:/temp/testGrids/edge_twi.tif"
```

The parameter `path` is essentially the directory location where the raster files are sitting. If needed, we may also want to do recursive listing into directories that are within that path directory. We want `list.files()` to return all the files (in our case) that have the .tif extension. This criteria is set via the `pattern` parameter, such that $ at the end means that this is end of string, and but adding \\. ensures that you match only files with extension .tif, otherwise it may list (if they exist), files that end in .atif as an example. You may guess that any other type of pattern matching criteria could be used to suit your own specific data. The `full.names` logical parameter is just a question of whether we want to return the full pathway address of the raster file, in which case, we do.

All we then need to do is perform a raster stack of these individual rasters, then perform the intersection. This is really the handy feature where to perform the stack, we still need not require the loading of the rasters into the R memory—they are still on file!.

```
files <- list.files(path = "C:/temp/testGrids/",
    pattern = "\\.tif$", full.names = TRUE)

# stack rasters
r1 <- raster(files[1])
for (i in 2:length(files)) {
    r1 <- stack(r1, files[i])
}
r1

## class         : RasterStack
## dimensions    : 400, 577, 230800, 5  (nrow, ncol, ncell, nlayers)
## resolution    : 90, 90  (x, y)
## extent        : 738698.6, 790628.6, 6643808, 6679808
                  (xmin, xmax, ymin, ymax)
## coord. ref. : +proj=utm +zone=55 +south +datum=WGS84
## +units=m +no_defs +ellps=WGS84 +towgs84=0,0,0
## names         : edge_elevation, edge_landsat_b3,
                  edge_landsat_b4, edge_radK, edge_twi
## min values    :     181.420395,        18.864470,
##                     13.184220, -0.009290, 9.801188
## max values    :     960.10742,        170.51700,
##                     154.57581,   5.16667, 23.89634
```

Note that the stacking of rasters can only be possible if they are all equivalent in terms of resolution and extent. If they are not you will find the other raster package functions `resample` and `projectRaster` as invaluable methods for harmonizing all your different raster layers. With the stacked rasters, we can perform the soil point data intersection as done previously.

```
DSM_data <- extract(r1, edgeroi_splineCarbon, sp = 1,
method = "simple")
```

4.3 Some Exploratory Data Analysis

We will continue using the `DSM_data` object that was created in the previous section. As the data set was saved to file you will also find it in your working directory. Type `getwd()` in the console to indicate the specific file location. So lets read the file in using the `read.table` function:

```
edge.dat <- read.table("edgeroiSoilCovariates_C.TXT",
sep = ",", header = T)
str(edge.dat)

## 'data.frame': 341 obs. of  15 variables:
##  $ id           : Factor w/ 341 levels "ed001",
                    "ed002",..: 1 2 3 4 ...
```

```
## $ east        : int  741912 744662 747412 750212
                   752912 755712 ...
## $ north       : int  6678083 6677983 6677933
                   6677933 6677883 ...
## $ X0.5.cm     : num  28.8 23.7 12.5 7.7 7.5 8.7 17.3
                   11.7 10.2 13.1 ...
## $ X5.15.cm    : num  18.6 14.7 9.3 6.3 6.9 7.8 12.6
                   10.4 7.7 10.5 ...
## $ X15.30.cm   : num  10.5 6.5 6.6 5.5 6.5 6.6 8 8.7 5.7 8 ...
## $ X30.60.cm   : num  9.2 6.6 5.6 5.6 6 5.9 7.3 6.6 5.7 7.2 ...
## $ X60.100.cm  : num  5.3 6.1 4 4.1 4.4 4.8 3.9 3.3 4.5 4.4 ...
## $ X100.200.cm: num  3.1 3.3 1.2 1.8 2.7 1.9 1.3
                   2.4 1.7 0.8 ...
## $ soil.depth  : int  260 260 260 260 260 259 260
                   260 253 260 ...
## $ elevation   : num  186 187 192 193 197 ...
## $ twi         : num  22.9 23.5 23.1 22.8 22.2 ...
## $ radK        : num  1.122 0.983 0.918 0.954 0.784 ...
## $ landsat_b3  : num  62.3 59.6 67.3 57.9 49 ...
## $ landsat_b4  : num  54.9 51.7 56.1 46.6 39.2 ...
```

Hereafter soil carbon density will be referred to as SOC for simplicity. Now lets firstly look at some of the summary statistics of SOC (we will just concentrate on the 0–5 cm depth interval in these following examples).

```
round(summary(edge.dat$X0.5.cm), 1)
```

```
##    Min. 1st Qu.  Median    Mean 3rd Qu.    Max.
##     0.3    11.9    16.4    18.9    23.1    93.1
```

The observation that the mean and median are not equivalent indicates the distribution of this data deviates from normal. To assess this more formally, we can perform other analyses such as tests of skewness, kurtosis and normality. Here we need to use functions from the `fBasics` and `nortest` packages (If you do not have these already you should install them.)

```
library(fBasics)
library(nortest)
# skewness
sampleSKEW(edge.dat$X0.5.cm)
```

```
##      SKEW
## 0.1964286
```

```
# kurtosis
sampleKURT(edge.dat$X0.5.cm)
```

```
##     KURT
## 1.303571
```

Here we see that the data is positively skewed. A formal test for normality of the Anderson-Darling Test statistic. There are others, so its worth a look at the help files associated with the `nortest` package.

```
ad.test(edge.dat$X0.5.cm)

##
##  Anderson-Darling normality test
##
## data:  edge.dat$X0.5.cm
## A = 10.594, p-value < 2.2e-16
```

For this data to be normally distributed the p value should be > than 0.05. This is confirmed when we look at the histogram and qq-plot of this data in Fig. 4.4.

```
par(mfrow = c(1, 2))
hist(edge.dat$X0.5.cm)
qqnorm(edge.dat$X0.5.cm, plot.it = TRUE, pch = 4, cex = 0.7)
qqline(edge.dat$X0.5.cm, col = "red", lwd = 2)
```

The histogram on Fig. 4.4 shows that there are just a few high values that are more-or-less outliers in the data. Generally for fitting most statistical models, we need to assume our data is normally distributed. A way to make the data to be more normal is to transform it. Common transformations include the square root, logarithmic, or power transformations. Below is an example of taking the natural log transform of the data.

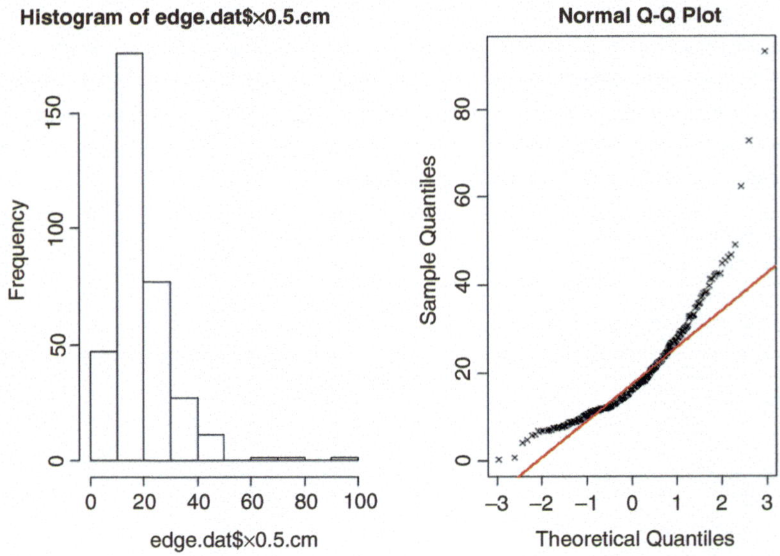

Fig. 4.4 Histogram and qq-plot of SOC in the 0–5 cm depth interval

```
sampleSKEW(log(edge.dat$X0.5.cm))

##        SKEW
## 0.03287885

sampleKURT(log(edge.dat$X0.5.cm))

##      KURT
## 1.196472

ad.test(log(edge.dat$X0.5.cm))

##
##  Anderson-Darling normality test
##
## data:  log(edge.dat$X0.5.cm)
## A = 1.9117, p-value = 6.935e-05
```

While not perfect, this is an improvement from before. This is also apparent when we do the plots shown on Fig. 4.5.

```
par(mfrow = c(1, 2))
hist(log(edge.dat$X0.5.cm))
qqnorm(log(edge.dat$X0.5.cm), plot.it = TRUE, pch = 4, cex = 0.7)
qqline(log(edge.dat$X0.5.cm), col = "red", lwd = 2)
```

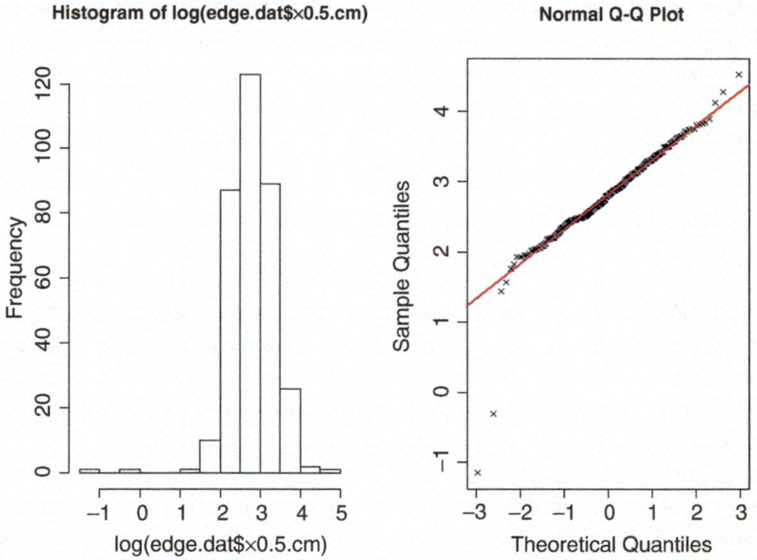

Fig. 4.5 Histogram and qq-plot of the natural log of SOC in the 0–5 cm depth interval

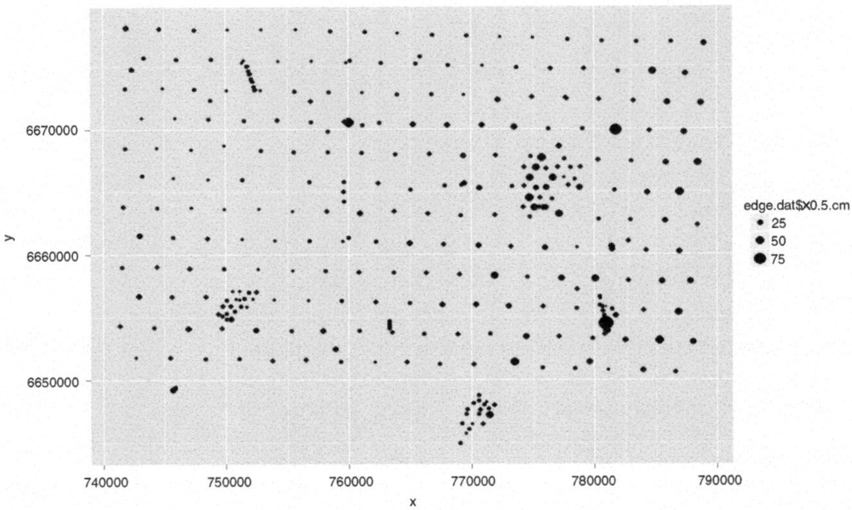

Fig. 4.6 Spatial distribution of points in the Edgeroi for the untransformed SOC data at the 0–5 cm depth interval

We could investigate other data transformations or even investigate the possibility of removing outliers or some such extraneous data, but generally we can be pretty satisfied with working with this data from a statistical viewpoint.

Another useful exploratory test is to visualize the data in its spatial context. Mapping the point locations with respect to the target variable by either altering the size or color of the marker gives a quick way to examine the target soil attribute spatial variability. Using the `ggplot2` package, we could create the plot as shown in Fig. 4.6.

```
library(ggplot2)
ggplot(edge.dat, aes(x = x, y = y)) +
geom_point(aes(size = edge.dat$X0.5.cm))
```

On Fig. 4.6 (which illustrates the untransformed data), there is a subtle east to west trend of high to low values. This trend is generally related to differences in land use in this area where intensive cropping is practiced in the western area where the land is an open floodplain. To the east the land is slightly undulating and land use is generally associated with pastures and natural vegetation.

Ultimately we are interested in making maps. So, as a first exercise and to get a clearer sense of the "spatial structure" of the data it is good to use some interpolation method to estimate SOC values at all of the unvisited locations. A couple of ways of doing this is the inverse distance weighted (IDW) interpolation and kriging.

For IDW predictions at unvisited locations are calculated as a weighted average of the values available at the known points, where the weights are based only by distance from the interpolation location. Kriging is a similar distance weighted interpolation method based on values at observed locations, except it has an underlying model of the spatial variation of the data. This model is a variogram which describes the auto-correlation structure of the data as a function of distances. Kriging is usually superior to other means of interpolation because it provides an optimal interpolation estimate for a given coordinate location, as well as a prediction variance estimate. Kriging is very popular in soil science, and there are many variants of it. For further information and theoretical underpinnings of kriging or other associated geostatistical methods, with special emphasis for the soil sciences it is worth consulting Webster and Oliver (2001).

Functions for IDW interpolation and kriging are found in the `gstat` package. To initiate these interpolation methods, we first need to prepare a grid of points upon which the interpolators will be used. This can be done by extracting the coordinates from either of the 90 m resolution rasters we have for the Edgeroi. As will be seen later, this step can be made redundant because we can actually interpolate directly to raster. Nevertheless, the extract the pixel point coordinates from a raster we do the following using the `elevation` raster from `edgeroiCovariates`. (Make sure both `raster` and `ithir` packages are loaded).

```
data(edgeroiCovariates)

tempD <- data.frame(cellNos = seq(1:ncell(elevation)))
tempD$vals <- getValues(elevation)
tempD <- tempD[complete.cases(tempD), ]
cellNos <- c(tempD$cellNos)
gXY <- data.frame(xyFromCell(elevation, cellNos, spatial = FALSE))
```

The script above essentially gets the pixels which have values associated with them (discards all NA occurrences), and then uses the cell numbers to extract the associated spatial coordinate locations using the `xyFromCell` function. The result is saved in the `gXY` object.

Using the `idw` function from `gstat` we fit the formula as below. We need to specify the observed data, their spatial locations, and the spatial locations of the points we want to interpolate onto. The `idp` parameter allows you to specify the inverse distance weighting power. The default is 2, yet can be adjusted if you want to give more weighting to points closer to the interpolation point. As we can not evaluate the uncertainty of prediction with IDW, we can not really optimize this parameter.

```
library(gstat)
names(edge.dat)[2:3] <- c("x", "y")
IDW.pred <- idw(log(edge.dat$X0.5.cm) ~ 1, locations = ~x + y,
data = edge.dat, newdata = gXY, idp = 2)

## [inverse distance weighted interpolation]
```

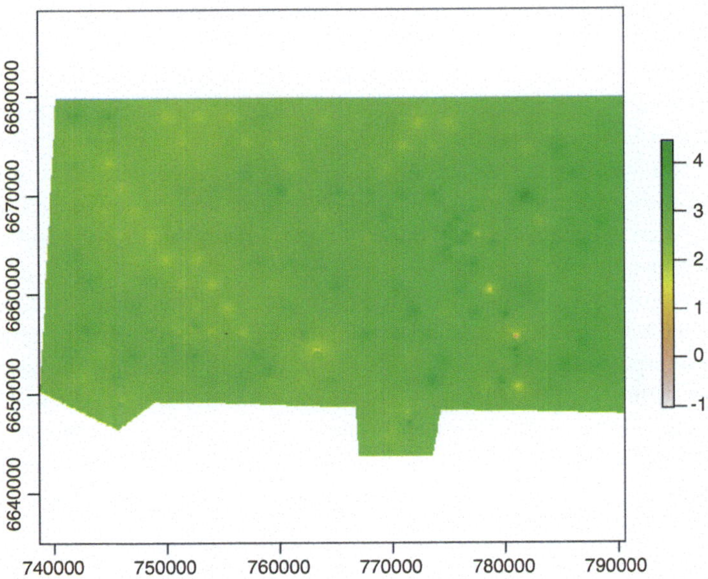

Fig. 4.7 Map of log SOC (0–5 cm) predicted using IDW

Plotting the resulting map (Fig. 4.7) can be done using the following script.

```
IDW.raster.p <- rasterFromXYZ(as.data.frame(IDW.pred[, 1:3]))
plot(IDW.raster.p)
```

For soil science it is more common to use kriging for the reasons that we are able to formally define the spatial relationships in our data and get an estimate of the prediction uncertainty. As mentioned before this is done using a variogram. Variograms measure the spatial auto-correlation of phenomena such as soil properties (Pringle and McBratney 1999). The average variance between any pair of sampling points (calculated as the semi-variance) for a soil property S at any point of distance h apart can be estimated by the formula:

$$\gamma(h) = \frac{1}{2m(h)} \sum_{i=1}^{m(h)} \{s(x_i) - s(x_i + h)\}^2 \tag{4.1}$$

where $\gamma(h)$ is the average semi-variance, m is the number of pairs of sampling points, s is the value of the attribute under investigation, x are the coordinates of the point, and h is the lag (separation distance of point pairs). Therefore, in accordance with the "law of geography", points closer together will show smaller semi-variance (higher correlation), whereas pairs of points farther away from each other will display larger semi-variance. A variogram is generated by plotting the average semi-variance against the lag distance. Various models can be fitted to this empirical variogram where four of the more common ones are the linear model, the spherical model, the exponential model, and the Gaussian model. Once an appropriate variogram has been modeled it is then used for distance weighted interpolation (kriging) at unvisited locations.

First, we calculate the empirical variogram i.e calculate the semivariances of all point pairs in our data set. Then we fit a variogram model (in this case we will use a spherical model). To do this we need to make some initial estimates of this models parameters; namely, the nugget, sill, and range. The nugget is the very short-range error (effectively zero distance) which is often attributed to measurement errors. The sill is the limit of the variogram (effectively the total variance of the data). The range is the distance at which the data are no longer auto-correlated. Once we have made the first estimates of these parameters, we use the `fit.variogram` function for their optimization. The `width` parameter of the variogram function is the width of distance intervals into which data point pairs are grouped or binned for semi variance estimates as a function of distance. An automated way of estimating the variogram parameters is to use the `autofitVariogram` function from the `automap` package. For now we will stick with the `gstat` implementation.

```
vgm1 <- variogram(log(X0.5.cm) ~ 1, ~x + y, edge.dat, width = 400,
cressie = TRUE, cutoff = 10000)
mod <- vgm(psill = var(log(edge.dat$X0.5.cm)),
"Exp", range = 5000, nugget = 0)
model_1 <- fit.variogram(vgm1, mod)
model_1

##   model     psill   range
## 1   Nug 0.1003208   0.000
## 2   Exp 0.0800328 988.807
```

The plot in Fig. 4.8 shows both the empirical variogram together with the fitted variogram model line.

```
plot(vgm1, model = model_1)
```

The variogram is indicating there is a relative high degree of nugget variation compared to the sill variation. There is some spatial in the data up to around 1000 m. Ultimately this will mean that there will be quantifiable prediction variances. To demonstrate this, lets perform the kriging to make a map, but more importantly look at the variances associated with the predictions. Here we use the `krige` function, which is not unlike using `idw` function, except that we have the variogram model parameters as additional information.

```
krig.pred <- krige(log(edge.dat$X0.5.cm) ~ 1, locations = ~x + y,
data = edge.dat, newdata = gXY, model = model_1)
```

We can make the maps as we did before, but now we can also look at the variances of the predictions too (Fig. 4.9).

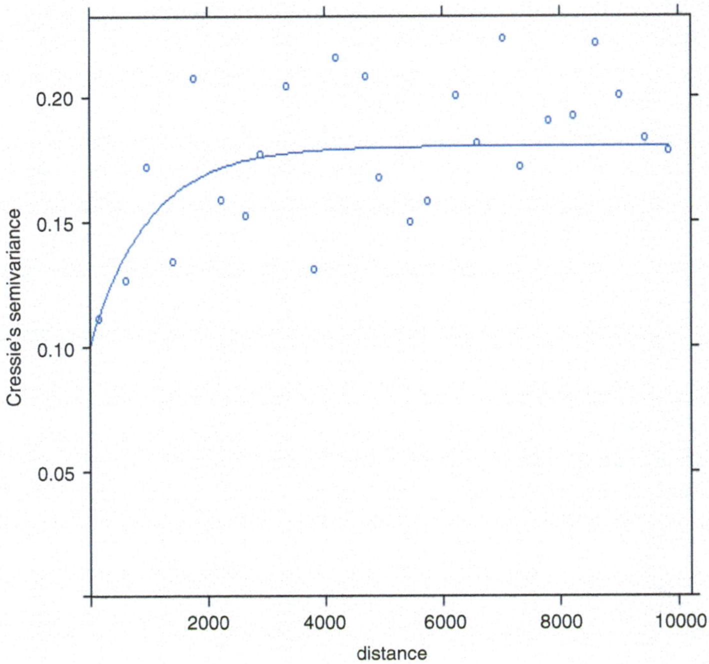

Fig. 4.8 Empirical variogram and spherical variogram model of log SOC for the 0–5 cm depth interval

```
par(mfrow = c(2, 1))
krig.raster.p <- rasterFromXYZ(as.data.frame(krig.pred[, 1:3]))
krig.raster.var <- rasterFromXYZ(as.data.frame
(krig.pred[, c(1:2, 4)]))
plot(krig.raster.p, main = "ordinary kriging predictions")
plot(krig.raster.var, main = "ordinary kriging variance")
```

Understanding the geostatisitcal properties of the target soil variable of interest is useful in its own right. However, it is also important to determine whether there is further spatial relationships in the data that can be modeled with environmental covariate information. Better still is to combine both spatial model approaches together (more of which will be discussed later on about this).

Ideally when we want to predict soil variables using covariate information is that there is a reasonable correlation between them. We can quickly assess these using the base `cor` function, for which we have used previously.

```
edge.dat$logC_0_5 <- log(edge.dat$X0.5.cm)
names(edge.dat)

## [1] "id"          "x"           "y"           "X0.5.cm"
## [5] "X5.15.cm"    "X15.30.cm"   "X30.60.cm"   "X60.100.cm"
```

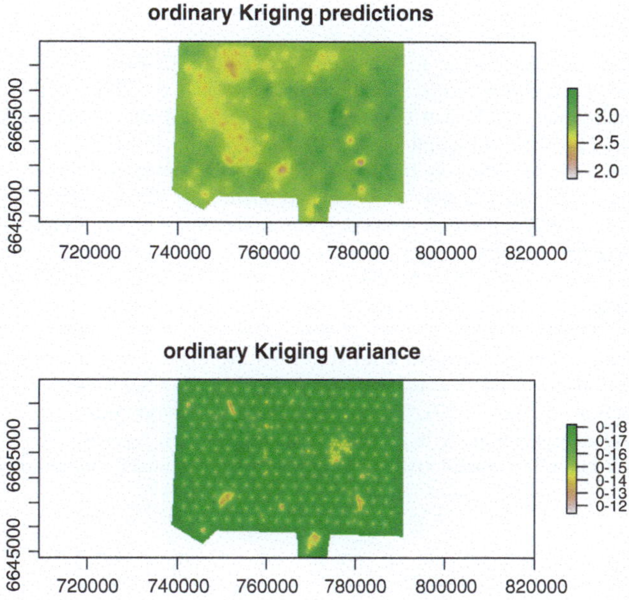

Fig. 4.9 Kriging predictions and variances for log SOC (0–5 cm)

```
##  [9] "X100.200.cm" "soil.depth"  "elevation"   "twi"
## [13] "radK"        "landsat_b3"  "landsat_b4"  "logC_0_5"
```

```
cor(edge.dat[, c("elevation", "twi", "radK",
    "landsat_b3", "landsat_b4")], edge.dat[, "logC_0_5"])
```

```
##                    [,1]
## elevation   0.440924269
## twi        -0.408020193
## radK        0.094710772
## landsat_b3 -0.002556606
## landsat_b4  0.060669516
```

It appears the highest correlations with log SOC are the variables derived from the digital elevation model: `elevation` and `twi`. Weak correlation is found for the other covariates. The following chapter of the book we will explore a range of models for mapping the soil as a function of this suite of covariates.

References

Arrouays D, McKenzie N, Hempel J, Richer de Forges A, McBratney A (eds) (2014) GlobalSoilMap: basis of the global spatial soil information system. CRC Press, Leiden

Bishop TFA, McBratney AB, Laslett GM (1999) Modelling soil attribute depth functions with equal-area quadratic smoothing splines. Geoderma 91:27–45

Malone BP, McBratney AB, Minasny B, Laslett GM (2009) Mapping continuous depth functions of soil carbon storage and available water capacity. Geoderma 154:138–152

McBratney AB, Ward WT, McBratney AB (1989) Soil studies in the Lower Namoi Valley: methods and data. The Edgeroi data set, 2 vols. CSIRO Division of Soils, Adelaide

Minasny B, McBratney AB (2010) Methodologies for Global Soil mapping. In: Digital soil mapping: bridging research, environmental application, and operation, chapter 34. Springer, Dordrecht, pp 429–425

Ponce-Hernandez R, Marriott FHC, Beckett PHT (1986) An improved method for reconstructing a soil profile from analysis of a small number of samples. J Soil Sci 37:455–467

Pringle MJ, McBratney AB (1999) Estimating average and proportional variograms of soil properties and their potential use in precision agriculture. Precis Agric 1:125–152

Webster R, Oliver MA (2001) Geostatistics for environmental scientists. John Wiley and Sons Ltd, West Sussex

Chapter 5
Continuous Soil Attribute Modeling and Mapping

The implementation of some of the most commonly used model functions used for digital soil mapping will be covered in this chapter. Before this is done however, some general concepts of model validation are covered.

5.1 Model Validation

Essentially, whenever we train or calibrate a model, we can then generate some predictions. The question one needs to ask is how good are those predictions? Generally, we confront this question by comparing observed values with their corresponding predictions. Some of the more common "quality" measures are the root mean square error (RMSE), bias, coefficient of determination or commonly the R^2 value, and concordance. You will also find in the digital soil mapping and general statistical literature various other model assessment statistics. The RMSE is defined as:

$$RMSE = \sqrt[2]{\left(\frac{\sum_{i=1}^{n}(obs_i - pred_i)^2}{n}\right)} \tag{5.1}$$

where *obs* is the observed soil property, *pred* is the predicted soil property from a given model, and n is the number of observations i. Bias, also called the mean error of prediction and is defined as:

$$bias = \frac{\sum_{i=1}^{n} pred_i - obs_i}{n} \tag{5.2}$$

The R^2 is evaluated as the square of the sample correlation coefficient (Pearson's) between the observations and their corresponding predictions. Pearson's correlation

B.P. Malone et al., *Using R for Digital Soil Mapping*,
Progress in Soil Science, DOI 10.1007/978-3-319-44327-0_5

coefficient r when applied to observed and predicted values is defined as:

$$r = \frac{\sum_{i=1}^{n}(obs_i - \overline{obs})(pred_i - \overline{pred})}{\sqrt[2]{\sum_{i=1}^{n}(obs_i - \overline{obs})^2}\sqrt[2]{\sum_{i=1}^{n}(pred_i - \overline{pred})^2}} \tag{5.3}$$

The R^2 measures the precision of the relationship (between observed and predicted). Concordance, or more formally—Lin's concordance correlation coefficient (Lin 1989), on the other hand is a single statistic that both evaluates the accuracy and precision of the relationship. It is often referred to as the goodness of fit along a 45 degree line. Thus it is probably a more useful statistic than the R^2 alone. Concordance ρ_c is defined as:

$$\rho_c = \frac{2\rho\sigma_{pred}\sigma_{obs}}{\sigma_{pred}^2 + \sigma_{obs}^2 + (\mu_{pred} - \mu_{obs})^2} \tag{5.4}$$

where μ_{pred} and μ_{obs} are the means of the predicted and observed values respectively. σ_{pred}^2 and σ_{obs}^2 are the corresponding variances. ρ is the correlation coefficient between the predictions and observations.

5.1.1 Model Goodness of Fit

So lets fit a simple linear model. We will use the `soil.data` set used before in the introductory to `R` chapter. First load the data in. We then want to regress CEC content on clay (also be sure to remove as NAs).

```
library(ithir)
library(MASS)
data(USYD_soil1)
soil.data <- USYD_soil1
mod.data <- na.omit(soil.data[, c("clay", "CEC")])
mod.1 <- lm(CEC ~ clay, data = mod.data, y = TRUE, x = TRUE)
mod.1

##
## Call:
## lm(formula = CEC ~ clay, data = mod.data, x = TRUE, y = TRUE)
##
## Coefficients:
## (Intercept)         clay
##      3.7791       0.2053
```

You will recall that this is the same model that we fitted during the introduction to `R` chapter. What we now want to do is evaluate some of the model quality statistics that were just described. Conveniently, these are available in the `goof` function in the `ithir` package. We will use this function a lot during this chapter, so it might be useful to describe it. `goof` takes four inputs. A vector of `observed` values, a vector of `predicted` values, a logical choice of whether an output plot is required, and a character input of what type of output is required. There are number of possible goodness of fit statistics that can be requested, with only some being used frequently in digital soil mapping projects. Therefore setting the `type` parameter to `"DSM"` will output only the R^2, RMSE, MSE, bias and concordance statistics as these are most relevant to DSM. Additional statistics can be returned if `"spec"` is specified for the `type` parameter

```
goof(observed = mod.data$CEC, predicted = mod.1$fitted.values,
type = "DSM")
```

```
##          R2 concordance      MSE     RMSE bias
## 1 0.4173582   0.5888521 14.11304 3.756733    0
```

You may wish to generate a plot in which case you would set the `plot.it` logical to `TRUE`.

This model `mod.1` does not seem to be too bad. On average the predictions are 3.75 cmol (+)/kg off the true value. The model on average is neither over- or under-predictive, but we can see that a few high CEC values are influencing the concordance and R^2. This outcome may mean that there are other factors that influence the CEC, such as mineralogy type.

5.1.2 Model Validation

Above we performed goodness of fit assessment of the `mod.1` model. Usually it is more appropriate however to validate a model using data that was not included for model fitting. Model validation has a few different forms. For completely unbiased assessments of model quality it is ideal to have an additional data set that is completely independent of the model data. It is recommended that a design based random sampling from the target area be conducted, to which there are a few types such as simple random sampling and stratified simple random sampling. Further information regarding sampling, sampling designs, their formulation and the relative advantages and constraints of each are described in de Gruijter et al. (2006). Usually from an operational perspective it is difficult to arrange the additional costs of organising and implementing some sort of probability sampling for determining unbiased model quality assessment. The alternative is to perform some sort of data sub-setting, such that with a data set we split it into a set for model calibration and another set for validation. This type of procedure can take different forms: the two main ones being random-hold back and leave-one-out-cross-validation (LOCV). Random-hold back (or sometimes k-fold validation) is where we may sample a data set of some pre-determined proportion (say 70 %) for which is used for model calibration. We then validate the model using the other 30 % of the data. For k-fold validation we divide the data set into equal sized partitions or folds, with all but one of the folds being used for the model calibration, the remaining fold is used for validation. We could repeat this k-fold process a number of times, each time using a different random sample from the data set for model calibration and validation. This allows one to efficiently derive distributions of the validation statistics as a means of assessing the stability and sensitivity of the models and parameters.

LOCV involves a little more computation such that if we had *n* number of data, we would subset *n-1* of these data, and fit a model. Using this model we would make a prediction for the single data that was left out of the model (and save the residual). This is repeated for all *n*. LOCV would be undertaken when there are very few data to work with. When we can sacrifice a few data points, the random-hold back or k-fold cross-validation procedure would be acceptable.

When we are validating trained models with some sort of data sub-setting mechanism, always keep in mind that the validation statistics will be biased. As Brus et al. (2011) explains, the sampling from the target mapping area to be used for DSM is more often than not from legacy soil survey, to which would not have been based on a probability sampling design. Therefore, that sample will be biased i.e not a true representation of the total population. Even though we may randomly select observations from the legacy soil survey sites, those validation points do not become a probability sample of the target area, and consequently will only provide biased estimates of model quality. Thus an independent probability sample is required. Further ideas on the statistical validation of models can be found in Hastie et al. (2001).

So lets implement some of the validation techniques in `R`. We will use the same data as before i.e regressing CEC with clay content. First we will do the random-back validation using 70 % of the data for calibration. A random sample of the data will be performed using the `sample` function.

```
set.seed(123)
training <- sample(nrow(mod.data), 0.7 * nrow(mod.data))
training
```

```
##   [1]  42 115  59 127 134   7  74 125  77  63 131  62  91 138  14 118  32
##  [18]   6 146 122 113  87  80 123 124  86  66  71  35  18 112 104  79  90
##  [35]   3  54  84  24 136  25  16  44 105  38 106  15 109  47  27 110   5
##  [52]  43  76  12  52  19  93  68 114  33  58   9 139  23  67  37  65 143
##  [69] 135  34 121  48  53   1 108 102  98  95 100   8  17 145  70  50 141
##  [86]  64  60 140  92  10  82  36 142  72 120  57  40  96  83 107  28 101
```

These values correspond to row numbers which will correspond to the row which we will use for the calibration data. We subset these rows out of `mod.data` and fit a new linear model.

```
mod.rh <- lm(CEC ~ clay, data = mod.data[training, ],
y = TRUE, x = TRUE)
```

So lets evaluate the calibration model with `goof`:

```
goof(predicted = mod.rh$fitted.values,
    observed = mod.data$CEC[training])

##            R2 concordance      MSE     RMSE bias
## 1 0.4457907   0.6158071 12.31952 3.509917    0
```

But we are more interested in how this model performs when we use the validation data. Here we use the `predict` function to predict upon this data.

```
mod.rh.V <- predict(mod.rh, mod.data[-training, ])
goof(predicted = mod.rh.V, observed = mod.data$CEC[-training])

##            R2 concordance      MSE     RMSE       bias
## 1 0.3591283   0.5208349 18.35828 4.284656 -0.5355242
```

So the model is not as good as we first imagined.When we validate a model with an external data set, it is quite normal that the model will not perform nearly as well as when using calibration data. Set the `plot.it` parameter to `TRUE` and re-run the script above and you will see a plot like Fig. 5.1.

In fact the `mod.rh` model does not appear to perform too bad after all. A few of the high observed values contribute greatly to the validation diagnostics. A couple of methods are available to assess the sensitivity of these results. The first is to remove what could potentially be outliers from the data. The second is to perform

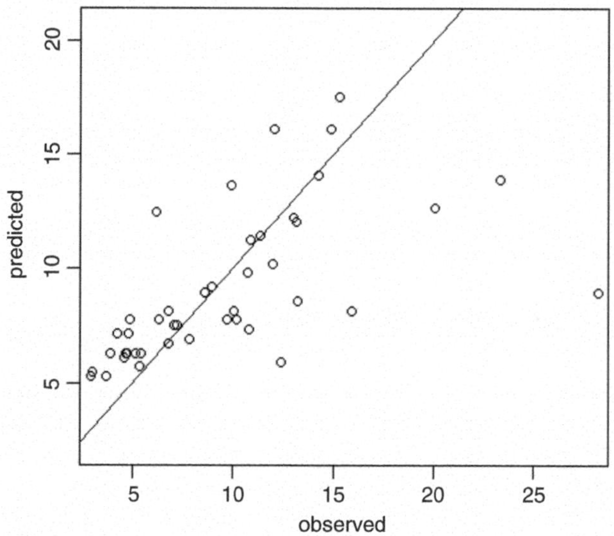

Fig. 5.1 Observed vs. predicted plot of CEC model (validation data set) with line of concordance (*red line*)

a sensitivity analysis such as bootstrapping where we iterate the data sub-setting procedure and evaluate the validation statistics each time to get a sense how much they vary.

At the most basic level, LOCV involves the use of a looping function or `for` loop. We have not really covered `for` loops yet, but essentially they can be used to great effect when we want to perform a particular analysis over-and-over. For example with LOCV, for each iteration or loop we take a subset of *n-1* rows and fit a model to them, then use that model to predict for the point left out of the calibration. Computationally it will look something like this:

```
looPred <- numeric(nrow(mod.data))
for (i in 1:nrow(mod.data)) {
    looModel <- lm(CEC ~ clay, data = mod.data[-i, ], y = TRUE,
    x = TRUE)
    looPred[i] <- predict(looModel, newdata = mod.data[i, ])
}
```

The `i` here is the counter, so for each loop it increases by 1 until we get to the end of the data set. As you can see, we can index the `mod.data` using the `i`, meaning that for each loop we will have selected a different calibration set. On each loop, the prediction on the point left out of the calibration is made onto the corresponding row position of the `looPred` object. Again we can assess the performance of the LOCV using the `goof` function.

```
goof(predicted = looPred, observed = mod.data$CEC)

##          R2 concordance      MSE     RMSE        bias
## 1 0.4025255   0.5790589 14.47653 3.804804 0.005758669
```

LOCV will generally be less sensitive to outliers, so overall these external validation results are not too different to those when we performed the internal validation. Make a plot of the LOCV results to visually compare against the internal validation.

5.2 Multiple Linear Regression

Multiple linear regression (MLR) is where we regress a target variable against more than one covariate. In terms of soil spatial prediction functions, MLR is a least-squares model whereby we want to predict a continuous soil variable from a suite of covariates. There are a couple of ways to go about this. We could just put everything (all the covariates) in the model and then fit it (estimate the model parameters). We could perform a stepwise regression model where we only enter variables that are statistically significant, based on some selection criteria. Alternatively we could fit what could be termed, an "expert" model, such that based on some pre-determined knowledge of the soil variable we are trying to model, we include covariates that best describe this knowledge. In some ways this is a biased model because we really don't know everything about (the spatial characteristics) the soil property under investigation. Yet in many situations it is better to rely on expert knowledge that is gained in the field as opposed to some other form.

So lets firstly get the data organized. Recall from before in the data preparatory exercises that we were working with the soil point data and environmental covariates for the Edgeroi area. These data are stored in the `edgeroi_splineCarbon` and `edgeroiCovariates` objects from the `ithir` package. For the succession of models to be used, we will concentrate on modelling and mapping the soil carbon stocks for the 0–5 cm depth interval. To refresh, lets load the data in, perform a log-transform of the soil carbon data (in order to make the distribution exhibit normality), then intersect the data with the available covariates.

```
library(ithir)
library(raster)
library(rgdal)
# point data
data(edgeroi_splineCarbon)
names(edgeroi_splineCarbon)[2:3] <- c("x", "y")
# natural log transform
edgeroi_splineCarbon$log_cStock0_5 <- log(edgeroi_splineCarbon
$X0.5.cm)

# grids
data(edgeroiCovariates)
```

Perform the covariate intersection.

```
coordinates(edgeroi_splineCarbon) <- ~x + y
# stack the rasters
covStack <- stack(elevation, twi, radK, landsat_b3, landsat_b4)
# extract
DSM_data <- extract(covStack, edgeroi_splineCarbon, sp = 1,
method = "simple")
DSM_data <- as.data.frame(DSM_data)
str(DSM_data)

## 'data.frame': 341 obs. of  16 variables:
##  $ id            : Factor w/ 341 levels "ed001","ed002",..:
                     1 2 3 4 5 ...
##  $ x             : num  741912 744662 747412 750212 752912 ...
##  $ y             : num  6678083 6677983 6677933 6677933
                     6677883 ...
##  $ X0.5.cm       : num  28.78 23.68 12.45 7.69 7.51 ...
##  $ X5.15.cm      : num  18.62 14.67 9.3 6.35 6.94 ...
##  $ X15.30.cm     : num  10.47 6.48 6.6 5.49 6.47 ...
##  $ X30.60.cm     : num  9.22 6.65 5.59 5.58 6 ...
##  $ X60.100.cm    : num  5.31 6.13 3.99 4.08 4.41 ...
##  $ X100.200.cm   : num  3.13 3.31 1.18 1.85 2.72 ...
##  $ soil.depth    : int  260 260 260 260 260 259 260 260
                     253 260 ...
```

```
##  $ log_cStock0_5: num  3.36 3.16 2.52 2.04 2.02 ...
##  $ elevation    : num  186 187 192 193 197 ...
##  $ twi          : num  22.9 23.5 23.1 22.8 22.2 ...
##  $ radK         : num  1.122 0.983 0.918 0.954 0.784 ...
##  $ landsat_b3   : num  62.3 59.6 67.3 57.9 49 ...
##  $ landsat_b4   : num  54.9 51.7 56.1 46.6 39.2 ...
```

It is a general preference to progress with a data frame of just the data and covariates required for the modelling. In this case, we will subset the columns pertaining to the target variable `log_cStock0_5`, the covariates and for later on, the spatial coordinates.

```
DSM_data <- DSM_data[, c(2:3, 11:16)]
```

Often it is handy to check to see whether there are missing values both in the target variable and of the covariates. It is possible that a point location does not fit within the extent of the available covariates. In these cases the data should be excluded. A quick way to assess whether there are missing or NA values in the data is to use the `complete.cases` function.

```
which(!complete.cases(DSM_data))

## integer(0)

DSM_data <- DSM_data[complete.cases(DSM_data), ]
```

There do not appear to be any missing data as indicated by the `integer(0)` output above i.e there are zero rows with missing information.

With the soil point data prepared, lets fit a model with everything in it (all covariates) to get an idea of how to parametise the MLR models in R. Remember the soil variable we are making a model for is the natural log of SOC for the 0–5 cm depth interval.

```
edge.MLR.Full <- lm(log_cStock0_5 ~ elevation + twi +
radK + landsat_b3 + landsat_b4, data = DSM_data)
summary(edge.MLR.Full)

##
## Call:
## lm(formula = log_cStock0_5 ~ elevation + twi + radK +
##     landsat_b3 + landsat_b4, data = DSM_data)
##
## Residuals:
##     Min      1Q  Median      3Q     Max
## -3.8090 -0.2312 -0.0156  0.2590  1.4837
##
## Coefficients:
##              Estimate Std. Error t value Pr(>|t|)
## (Intercept) 1.3294707  1.1371723   1.169 0.243194
## elevation   0.0045601  0.0012970   3.516 0.000498 ***
```

```
## twi           -0.0004933  0.0393538  -0.013 0.990006
## radK           0.0452981  0.0583984   0.776 0.438489
## landsat_b3     0.0046184  0.0021209   2.178 0.030135 *
## landsat_b4     0.0013177  0.0020299   0.649 0.516692
## ---
## Signif. codes:  0 '***' 0.001 '**' 0.01 '*' 0.05 '.' 0.1 ' ' 1
##
## Residual standard error: 0.4974 on 335 degrees of freedom
## Multiple R-squared:  0.2119,Adjusted R-squared:  0.2002
## F-statistic: 18.02 on 5 and 335 DF,  p-value: 7.876e-16
```

From the summary output above, it seems only a few of the covariates are significant in describing the spatial variation of the target variable. To determine the most parsimonious model we could perform a step wise regression using the `step` function. With this function we can also specify what direction we want step wise algorithm to proceed.

```
edge.MLR.Step <- step(edge.MLR.Full, trace = 0, direction="both")
summary(edge.MLR.Step)
```

```
##
## Call:
## lm(formula = log_cStock0_5 ~ elevation + landsat_b3,
   data = DSM_data)
##
## Residuals:
##     Min      1Q  Median      3Q     Max
## -3.7556 -0.2325 -0.0122  0.2611  1.4594
##
## Coefficients:
##             Estimate Std. Error t value Pr(>|t|)
## (Intercept) 1.411781   0.181818   7.765 9.84e-14 ***
## elevation   0.004641   0.000491   9.454  < 2e-16 ***
## landsat_b3  0.004684   0.001993   2.350   0.0193 *
## ---
## Signif. codes:  0 '***' 0.001 '**' 0.01 '*' 0.05 '.' 0.1 ' ' 1
##
## Residual standard error: 0.4961 on 338 degrees of freedom
## Multiple R-squared:  0.2091,Adjusted R-squared:  0.2044
## F-statistic: 44.69 on 2 and 338 DF,  p-value: < 2.2e-16
```

Comparing the outputs of both the full and stepwise MLR models, there is very little difference in the model diagnostics such as the R^2. Both models explain about 20 % of variation of the target variable. Obviously the "full" model is more complex as it has more parameters than the "step" model. If we apply Occam's Razor, the "step" model is preferable.

As described earlier, it is more acceptable to test the performance of a model based upon an external validation. Lets fit a new model using the covariates selected in the step wise regression to a random subset of the available data. We will sample 70 % of the available rows for the model calibration data set.

```
set.seed(123)
training <- sample(nrow(DSM_data), 0.7 * nrow(DSM_data))
edge.MLR.rh <- lm(log_cStock0_5 ~ elevation + landsat_b3,
data = DSM_data[training,])

# calibration predictions
MLR.pred.rhC <- predict(edge.MLR.rh, DSM_data[training, ])

# validation predictions
MLR.pred.rhV <- predict(edge.MLR.rh, DSM_data[-training, ])
```

Now we can evaluate the test statistics of the calibration model using the `goof` function.

```
# calibration
goof(observed = DSM_data$log_cStock0_5[training], predicted
= MLR.pred.rhC)

##          R2 concordance      MSE      RMSE bias
## 1 0.1927168   0.3265533 0.261809 0.5116728    0

# validation
goof(observed = DSM_data$log_cStock0_5[-training], predicted
= MLR.pred.rhV)

##          R2 concordance       MSE      RMSE        bias
## 1 0.2267094   0.4065357 0.2054525 0.4532687 -0.07370045
```

In this situation the calibration model does not appear to be over fitting because the test statistics for the validation are equal to or better to those of the calibration data. While this is a good result, the prediction model performs only moderately well by the fact there is a noticeable deviation between observations and corresponding model predictions. Examining other candidate models is a way to try to improve upon this results.

5.2.1 Applying the Model Spatially

From a soil mapping perspective the important question to ask is: What does the map look like that results from a particular model? In practice this can be answered by applying the model parameters to the grids of the covariates that were used in the model. There are a few options on how to do this.

5.2.1.1 Covariate Table

The traditional has been to collate a grid table where there would be two columns for the coordinates followed by other columns for each of the available covariates that were sourced. This was seen as an efficient way to organize all the covariate data as it ensured that a common grid was used which also meant that all the covariates are of the same scale in terms of resolution and extent. We can simulate the covariate table approach using the `edgeroiCovariates` object as below.

```
data(edgeroiCovariates)
covStack <- stack(elevation, twi, radK, landsat_b3, landsat_b4)
tempD <- data.frame(cellNos = seq(1:ncell(covStack)))
vals <- as.data.frame(getValues(covStack))
tempD <- cbind(tempD, vals)
tempD <- tempD[complete.cases(tempD), ]
cellNos <- c(tempD$cellNos)
gXY <- data.frame(xyFromCell(covStack, cellNos, spatial = FALSE))
tempD <- cbind(gXY, tempD)
str(tempD)
```

```
## 'data.frame': 201313 obs. of  8 variables:
##  $ x         : num  740274 740364 740454 740544 740634 ...
##  $ y         : num  6679763 6679763 6679763 6679763 6679763 ...
##  $ cellNos   : int  18 19 20 21 22 23 24 25 26 27 ...
##  $ elevation : num  183 184 185 184 183 ...
##  $ twi       : num  22.2 22.2 22.2 22.3 22.4 ...
##  $ radK      : num  1.04 1.06 1.09 1.12 1.15 ...
##  $ landsat_b3: num  74.1 75.9 67.5 66 65.4 ...
##  $ landsat_b4: num  65.3 66.7 58.6 57 56.2 ...
```

The result shown above is that the covariate table contains 201313 rows and has 8 variables. It is always necessary to have the coordinate columns, but some saving of memory could be earned if only the required covariates are appended to the table. It will quickly become obvious however that the covariate table approach could be limiting when mapping extents get very large or the grid resolution of mapping decreases, or both.

With the covariate table arranged it then becomes a matter of using the MLR `predict` function.

```
map.MLR <- predict(edge.MLR.rh, newdata = tempD)
map.MLR <- cbind(data.frame(tempD[, c("x", "y")]), map.MLR)
```

Now we can rasterise the predictions for mapping (Fig. 5.2) and grid export. In the example below we set the CRS to WGS84 Zone 55 before exporting the raster file out as a Geotiff file.

```
map.MLR.r <- rasterFromXYZ(as.data.frame(map.MLR[, 1:3]))
plot(map.MLR.r, main = "MLR predicted log SOC stock (0-5cm)")
# set the projection
crs(map.MLR.r) <- "+proj=utm +zone=55 +south +ellps=WGS84
+datum=WGS84 +units=m +no_defs"
writeRaster(map.MLR.r, "cStock_0_5_MLR.tif", format = "GTiff",
datatype = "FLT4S", overwrite = TRUE)
# check woking directory for presence of raster
```

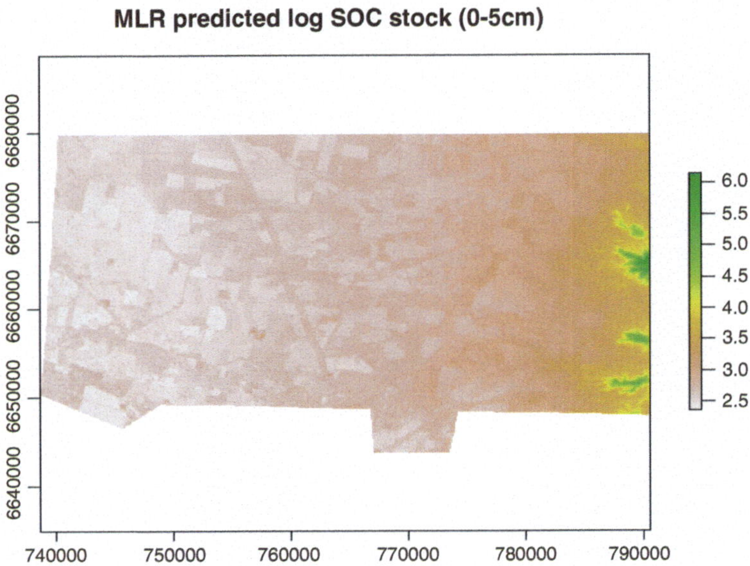

Fig. 5.2 MLR predicted log SOC stock 0–5 cm Edgeroi

Some of the parameters used within the `writeRaster` function that are worth noting include: `format`, which is the raster format that we want to write to. Here "GTiff" is being specified—use the `writeFormats` function to look at what other raster formats can be used. the parameter `datatype` is specified as "FLT4S" which indicates that a 4 byte, signed floating point values are to be written to file. Look at the function `dataType` to look at other alternatives, for example for categorical data where we may be interested in logical or integer values.

5.2.1.2 Raster Predictions

Probably a more efficient way of applying the fitted model is to apply it directly to the rasters themselves. This avoids the step of arranging all covariates into table format. If multiple rasters are being used, it is necessary to have them arranged as a `rasterStack` object. This is useful as it also ensures all the rasters are of the same extent and resolution. Here we can use the `raster predict` function such as below using the `covStack` raster stack as input.

```
map.MLR.r1 <- predict(covStack, edge.MLR.rh, "cStock_0_5_MLR.tif",
format = "GTiff", datatype = "FLT4S", overwrite = TRUE)
# check woking directory for presence of raster
```

The prediction function is quite versatile. For example we can also map the standard error of prediction or the confidence interval or the prediction interval even. The script below is an example of creating maps of the 90 % prediction intervals for the `edger.MLR.rh` model. We need to explicitly create a function called in this case `predfun` which will direct the raster `predict` function to output the predictions plus the upper and lower prediction limits. In the `predict` function we insert `predfun` for the `fun` parameter and control the output by changing the `index` value to either 1, 2, or 3 to request either the prediction, lower limit, upper limit respectively. Setting the `level` parameter to 0.90 indicates that we want to return the 90 % precision interval. The resulting plots are shown in Fig. 5.3.

```
par(mfrow = c(3, 1))
predfun <- function(model, data) {
v <- predict(model, data, interval = "prediction", level = 0.9)
}

map.MLR.r.low <- predict(covStack, edge.MLR.rh, "cStock_0_5_MLR_low.tif",
fun = predfun, index = 2, format = "GTiff", datatype = "FLT4S", overwrite = TRUE)
plot(map.MLR.r.low, main = "MLR predicted log SOC stock (0-5cm) lower limit")

map.MLR.r.pred <- predict(covStack, edge.MLR.rh, "cStock_0_5_MLR_pred.tif",
fun = predfun, index = 1, format = "GTiff", datatype = "FLT4S", overwrite = TRUE)
plot(map.MLR.r.pred, main = "MLR predicted log SOC stock (0-5cm)")

map.MLR.r.up <- predict(covStack, edge.MLR.rh, "cStock_0_5_MLR_up.tif",
fun = predfun, index = 3, format = "GTiff", datatype = "FLT4S", overwrite = TRUE)
plot(map.MLR.r.up, main = "MLR predicted log SOC stock (0-5cm) upper limit")
# check woking directory for presence of rasters
```

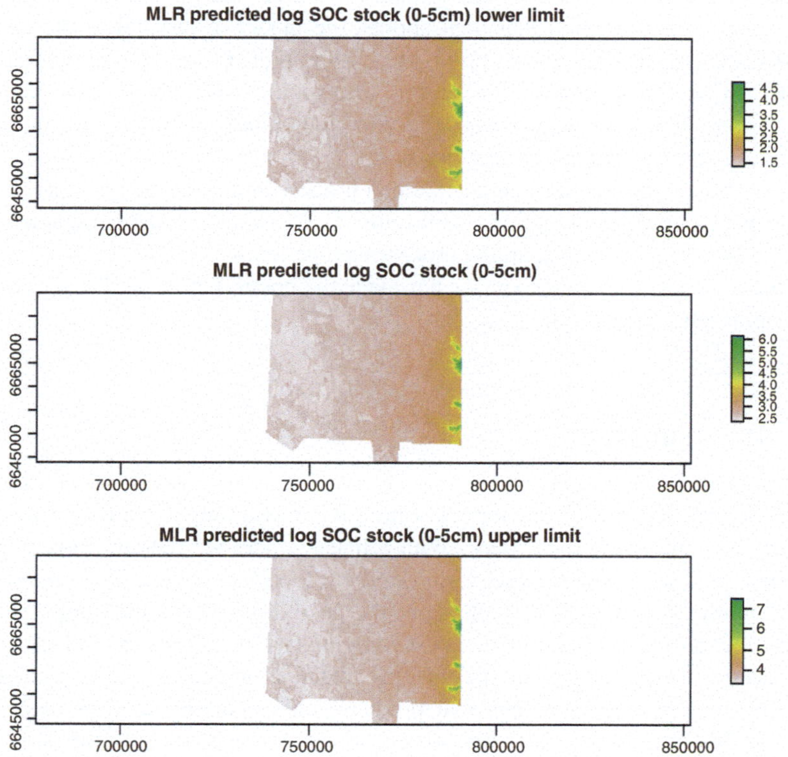

Fig. 5.3 MLR predicted log SOC stock 0–5 cm Edgeroi with lower and upper prediction limits

5.2.1.3 Directly to Rasters Using Parallel Processing

An extension of using the raster `predict` function is the apply the model again to the rasters, but to do it across multiple computer nodes. This is akin to breaking a job up into smaller pieces then processing the jobs in parallel rather than sequentially. The parallel component here is that the smaller pieces are passed to more than 1 compute nodes. Most desktop computers these days can have up to 8 compute nodes which can result in some excellent gains in efficiency when applying models across massive extents and or at fine resolutions. The `raster` package has some built in dependencies with other `R` packages that facilitate parallel processing options. For example the `raster` package ports with the `snow` package for setting up and controlling the compute node processes. The script below is an example of using 4 compute nodes to apply the `edge.MLR.rh` model to the `covStack` raster stack.

```
beginCluster(4)
cluserMLR.pred <- clusterR(covStack, predict, args = list(edge.MLR.rh),
filename = "cStock_0_5_MLR_pred.tif", format = "GTiff",
progress = FALSE, overwrite = T)
endCluster()
```

To set up the compute nodes, you use the `beginCluster` function and inside it, specify how many compute nodes you want to use. If empty brackets are used, the function will use 100 % of the compute resources. The `clusterR` function is the work horse function that then applies the model in parallel to the rasters. The parameters and subsequent options are similar to the raster `predict` function, although it would help to look at the help files on this function for more detailed explanations. It is always important after the prediction is completed to shutdown the nodes using the `endCluster` function.

The relative ease in setting up the parallel processing for our mapping needs has really opened up the potential for performing DSM using very large data sets and rasters. Moreover, using the parallel processing together with the file pointing ability (that was discussed earlier) `raster` has made the possibility of big DSM a reality, and importantly; practicable.

5.3 Decision Trees

Linear regression is a global model, where there is a single predictive formula holding over the entire data space. With a linear model we therefore make some assumptions about how our target variable relates to the covariates. These may often hold, however, it is models that allow one the flexibility of modelling non-linearity that are increasingly popular in the DSM community. One of these model structures are classification and regression trees (CART). These models are a non-parametric decision tree learning technique that produces either classification or regression trees. In this section we will concentrate on regression trees because our target variable is numeric i.e. a continuous variable. Later we will look at classification trees for categorical variables. Decision trees (either regression or classification) are formed by a collection of rules based on variables in the modeling data set:

- Rules based on variables' values are selected to get the best split to differentiate observations based on the dependent variable.
- Once a rule is selected and splits a node into two, the same process is applied to each subsequent node (i.e. it is a recursive procedure).
- Splitting stops when CART detects no further gain can be made, or some pre-set stopping rules are met. Alternatively, the data are split as much as possible and then the tree is later pruned.

Each branch of the tree ends in a terminal node. Each observation falls into one and exactly one terminal node, and each terminal node is uniquely defined by a set of rules. For a regression tree, the terminal node is a single value, or could be a regression model (which is the case for Cubist models which we will look at later). Implementation of regression trees in `R` is provided both through the `rpart` and `party` packages. We will use the `rpart` package and its `rpart` function. However, the `party` package through the `ctree` function offers more functionality, and implements the partitioning in a more statistical robust fashion. Both functions however can handle both continuous and categorical predictor variables.

Fitting a decision tree in `R` is quite similar to that for linear models:

```
library(rpart)
set.seed(123)
training <- sample(nrow(DSM_data), 0.7 * nrow(DSM_data))
edge.RT.Exp <- rpart(log_cStock0_5 ~ elevation + twi + radK +
landsat_b3 + landsat_b4, data = DSM_data[training, ],
control = rpart.control(minsplit = 50))
```

It is worthwhile to look at the help file for `rpart` particularly those aspects regarding the `rpart.control` parameters which control the `rpart` fit. Often it is helpful to just play around with the parameters to get a sense of what does what. Here for the `minsplit` parameter within `rpart.control` we are specifying that we want at least 50 observations in a node in order for a split to be attempted.

Detailed results of the model fit can be provided via the `summary` and `printcp` functions.

```
summary(edge.RT.Exp)
```

The `summary` output provides detailed information of the data splitting as well as information as to the relative importance of the covariates.

```
printcp(edge.RT.Exp)
```

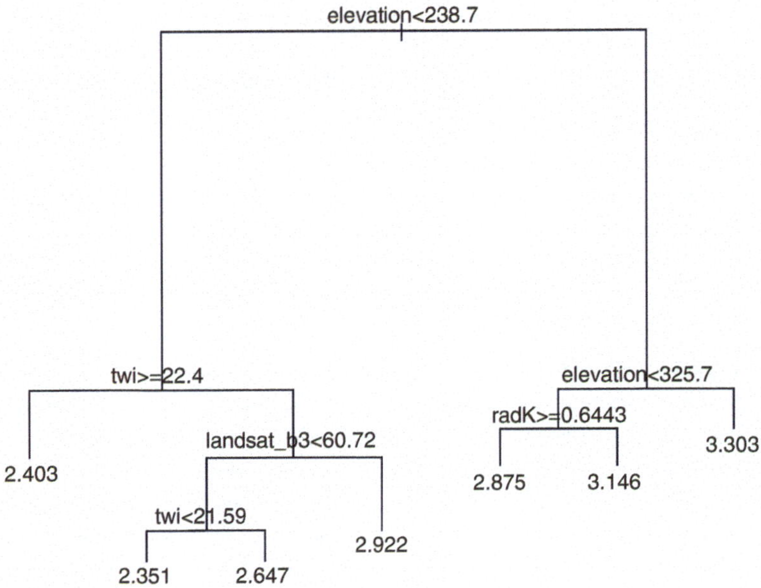

Fig. 5.4 Decision tree of log SOC stock 0–5 cm Edgeroi

The `printcp` function provides the useful output of indicating which covariates were included in the final model. For the visually inclined, a plot of the tree assists a lot to interpret the model diagnostics and assessing the important covariates too (Fig. 5.4).

```
plot(edge.RT.Exp)
text(edge.RT.Exp)
```

As before, we can use the `goof` function to test the performance of the model fit both internally and externally.

```
# Internal validation
RT.pred.C <- predict(edge.RT.Exp, DSM_data[training, ])
goof(observed = DSM_data$log_cStock0_5[training], predicted = RT.pred.C)

##          R2 concordance       MSE      RMSE          bias
## 1 0.2894225   0.4506149 0.2304465 0.4800485 -4.440892e-16

# External validation
RT.pred.V <- predict(edge.RT.Exp, DSM_data[-training, ])
goof(observed = DSM_data$log_cStock0_5[-training], predicted = RT.pred.V)

##          R2 concordance       MSE      RMSE        bias
## 1 0.2453704   0.4266996 0.1990041 0.4460987 -0.06496294
```

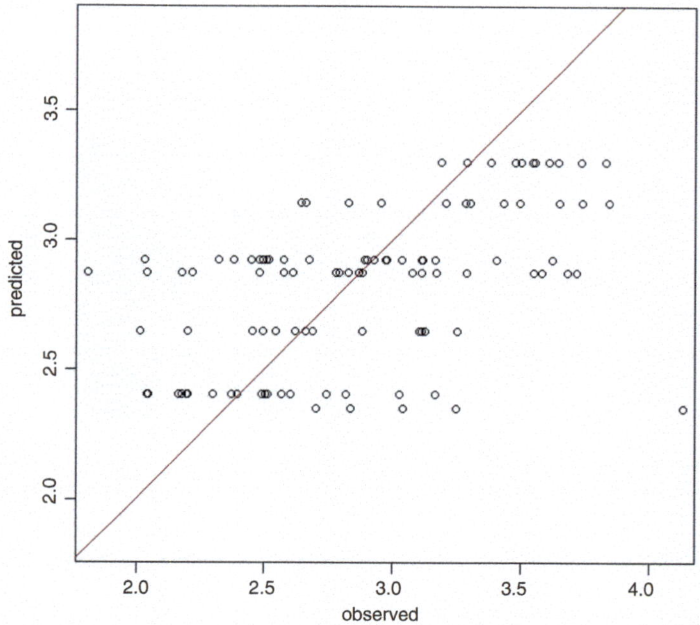

Fig. 5.5 Decision tree xy-plot plot of log SOC stock 0–5 cm (validation data set)

The decision tree model performance is not too dissimilar to the MLR model. Looking at the xy-plot from the external validation (Fig. 5.5) and the decision tree (Fig. 5.4), it becomes clear that a potential issue is apparent. This is: there are only a finite number of possible outcomes in terms of the predictions.

This finite property becomes obviously apparent once we make a map by applying the `edge.RT.Exp` model to the covariates (using the `raster` predict function and `covStack` object) (Fig. 5.6).

```
map.RT.r1 <- predict(covStack, edge.RT.Exp, "cStock_0_5_RT.tif",
format = "GTiff", datatype = "FLT4S", overwrite = TRUE)
```

```
plot(map.RT.r1, main = "Decision tree predicted 0-5cm log carbon stocks")
```

5.4 Cubist Models

The Cubist model is currently a very popular model structure used within the DSM community. Its popularity is due to its ability to "mine" non-linear relationships in data, but does not have the issues of finite predictions that occur for other decision and regression tree models. In similar vain to regression trees however, Cubist

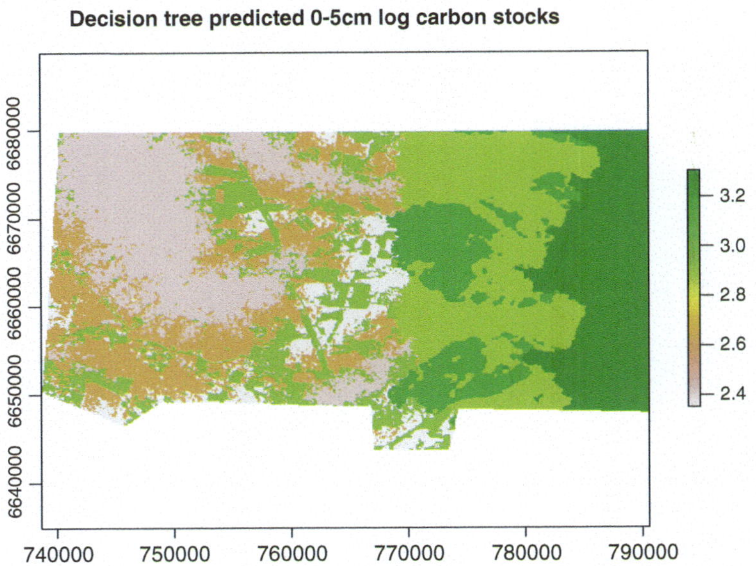

Fig. 5.6 Decision tree predicted log SOC stock 0–5 cm Edgeroi

models also are a data partitioning algorithm. The Cubist model is based on the M5 algorithm of Quinlan (1992), and is implemented in the `R` Cubist package.

The Cubist model first partitions the data into subsets within which their characteristics are similar with respect to the target variable and the covariates. A series of rules (a decision tree structure may also be defined if requested) defines the partitions, and these rules are arranged in a hierarchy. Each rule takes the form:

if [condition is true]
then [regress]
else [apply the next rule]

The condition may be a simple one based on one covariate or, more often, it comprises a number of covariates. If a condition results in being true then the next step is the prediction of the soil property of interest by ordinary least-squares regression from the covariates within that partition. If the condition is not true then the rule defines the next node in the tree, and the sequence of *if, then, else* is repeated. The result is that the regression equations, though general in form, are local to the partitions and their errors smaller than they would otherwise be. More details of the Cubist model can be found in the `Cubist` help files or Quinlan (1992).

Luckily, fitting a Cubist model in `R` is not too difficult—although it will be useful to spend some time playing around with many of the controllable parameters the function has. In the example below we can control the number of potential rules that could potentially partition the data (note this limits the number of possible rules, and does not necessarily mean that those number of rules will acutallu be realised i.e. the outcome is internally optimised). We can also limit the extrapolation of the model predictions, which is a useful model constraint feature. These various control parameters plus others can be adjusted within the `cubistControl` parameter. The committees parameter is specified as an integer of how many committee models (e.g.. boosting iterations) are required. Here we just set it to 1, but naturally it is possible to perform some sort of sensitivity analysis when this committee model option is set to greater than 1. In terms of specifying the target variable and covariates, we do not define a formula as we did earlier for the MLR model. Rather we specify the columns explicitly—those that are the target variable (`x`), and those that are the covariates (`y`).

```
library(Cubist)
set.seed(123)
training <- sample(nrow(DSM_data), 0.7 * nrow(DSM_data))
mDat <- DSM_data[training, ]

# fit the model
edge.cub.Exp <- cubist(x = mDat[, c("elevation", "twi", "radK", "landsat_b3",
"landsat_b4")], y = mDat$log_cStock0_5,
cubistControl(rules = 5, extrapolation = 5),committees = 1)
```

The output generated from fitting a Cubist model can be retrieved using the `summary` function. This provides information about the conditions for each rule, the regression models for each rule, and information about the diagnostics of the model fit, plus the frequency of which the covariates were used as conditions and/or within a model.

```
summary(edge.cub.Exp)

##
## Call:
## cubist.default(x = mDat[, c("elevation", "twi", "radK",
##  "landsat_b3", "landsat_b4")], y = mDat$log_cStock0_5, committees =
##  1, control = cubistControl(rules = 5, extrapolation = 5))
##
##
## Cubist [Release 2.07 GPL Edition]  Mon Feb 01 13:18:28 2016
## ---------------------------------
##
##     Target attribute 'outcome'
##
## Read 238 cases (6 attributes) from undefined.data
##
## Model:
##
##   Rule 1: [238 cases, mean 2.7634952, range -1.147828 to 4.533301,
##               est err 0.3358926]
##
##  outcome = -0.409619 + 0.0066 elevation + 0.063 twi + 0.0042 landsat_b4
##
##
## Evaluation on training data (238 cases):
##
##     Average  |error|           0.3968150
```

```
##     Relative |error|              0.99
##     Correlation coefficient       0.28
##
##
##  Attribute usage:
##    Conds  Model
##
##           100%    elevation
##           100%    twi
##           100%    landsat_b4
##
##
## Time: 0.0 secs
```

It appears the `edge.cub.Exp` model contains only 1 rule in this case. The useful feature of the Cubist model is that it does not unnecessarily overfit and partition the data. Lets see how well it validates.

```
# Internal validation
Cubist.pred.C <- predict(edge.cub.Exp, newdata = DSM_data[training, ])
goof(observed = DSM_data$log_cStock0_5[training], predicted = Cubist.pred.C)

##          R2 concordance       MSE      RMSE         bias
## 1 0.1752903   0.3173758 0.2678904 0.5175813 -0.009558701

# External validation
Cubist.pred.V <- predict(edge.cub.Exp, newdata = DSM_data[-training, ])
goof(observed = DSM_data$log_cStock0_5[-training], predicted = Cubist.pred.V)

##          R2 concordance       MSE      RMSE        bias
## 1 0.2297899   0.4131117 0.2084628 0.4565773 -0.09147446
```

The calibration model validates quite well, but its performance against the validation does not appear to be so good. From Fig. 5.7 it appears a few observation were very much under predicted which has had an impact on the subsequent model performance diagnostics, otherwise the Cubist model performs reasonably well.

Creating the map resulting from the `edge.cub.Exp` model can be implemented as before using the raster `predict` function (Fig. 5.8).

```
map.cubist.r1 <- predict(covStack, edge.cub.Exp, "cStock_0_5_cubist.tif",
format = "GTiff", datatype = "FLT4S", overwrite = TRUE)


plot(map.cubist.r1, main = "Cubist model predicted 0-5cm log carbon
stocks (0-5cm)")
```

5.5 Random Forests

An increasingly popular data mining algorithm in DSM and soil sciences, and even in applied sciences in general is the Random Forests model. This algorithm is provided in the `randomForest` package and can be used for both regression and

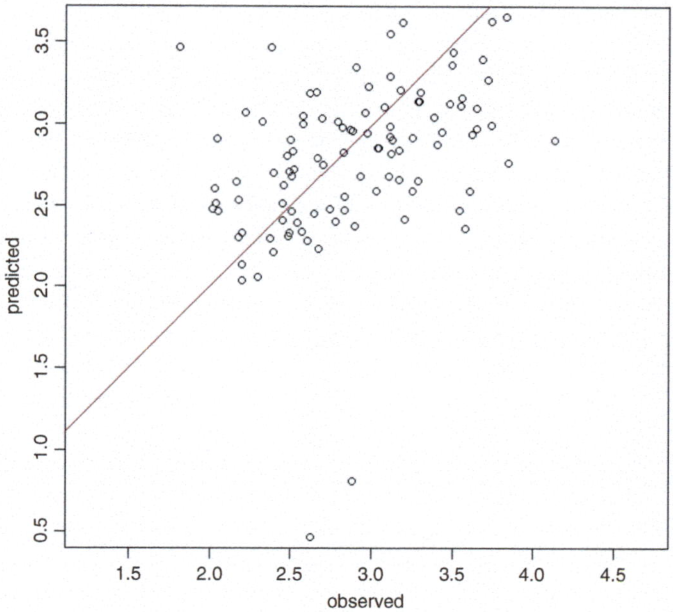

Fig. 5.7 Cubist model xy-plot plot of log SOC stock 0–5 cm (validation data set)

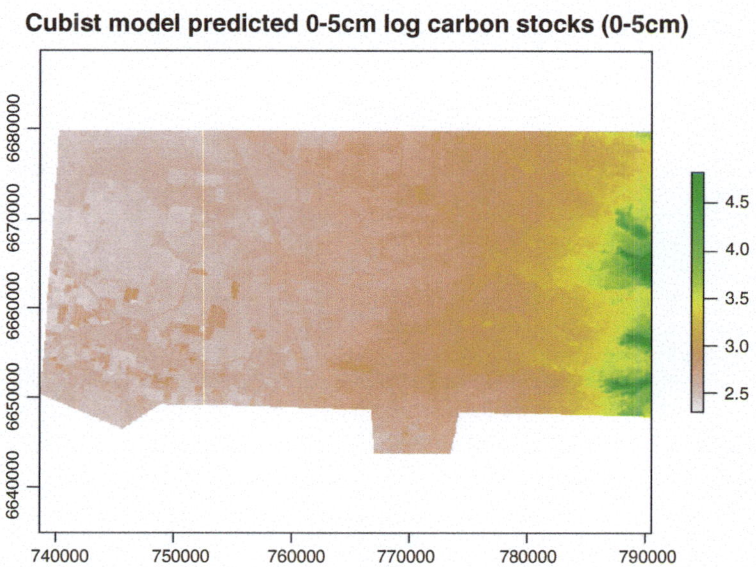

Fig. 5.8 Cubist model predicted log SOC stock 0–5 cm Edgeroi

classification purposes. Random Forests are a boosted decision tree model. Further, Random Forests are an ensemble learning method for classification (and regression) that operate by constructing a multitude of decision trees at training time, which are later aggregated to give one single prediction for each observation in a data set.For regression the prediction is the average of the individual tree outputs, whereas in classification the trees vote by majority on the correct classification (mode). For further information regarding Random Forest and their underlying theory it is worth consulting Breiman (2001) and Grimm et al. (2008) as an example of its application in DSM studies.

Fitting a Random Forest model in `R` is relatively straightforward. It is worth consulting the richly populated help files regarding the `randomForest` package and its functions. We will be using the `randomForest` function and a couple of extractor functions to tease out some of the model fitting diagnostics. Familiar will be the formula structure of the model. As for the Cubist model, there are many model fitting parameters to consider such as the number of trees to build (`ntree`) and the number of variables (covariates) that are randomly sampled as candidates at each decision tree split (`mtry`), plus many others. The `print` function allows one to quickly assess the model fit. The `importance` parameter (logical variable) used within the `randomForest` function specifies that we want to also assess the importance of the covariates used.

```
library(randomForest)
set.seed(123)
training <- sample(nrow(DSM_data), 0.7 * nrow(DSM_data))

# fit the model
edge.RF.Exp <- randomForest(log_cStock0_5 ~ elevation + twi + radK +
landsat_b3 + landsat_b4, data = DSM_data[training, ],
importance = TRUE, ntree = 1000)

print(edge.RF.Exp)

##
## Call:
##  randomForest(formula = log_cStock0_5 ~ elevation + twi + radK +
##   landsat_b3 + landsat_b4, data = DSM_data[training, ],
##   importance = TRUE,    ntree = 1000)
##                Type of random forest: regression
##                      Number of trees: 1000
## No. of variables tried at each split: 1
##
##           Mean of squared residuals: 0.2747959
##                     % Var explained: 15.62
```

Using the `varImpPlot` function allows one to visualize which covariates are of most importance to the prediction of our soil variable (Fig. 5.9).

```
varImpPlot(edge.RF.Exp)
```

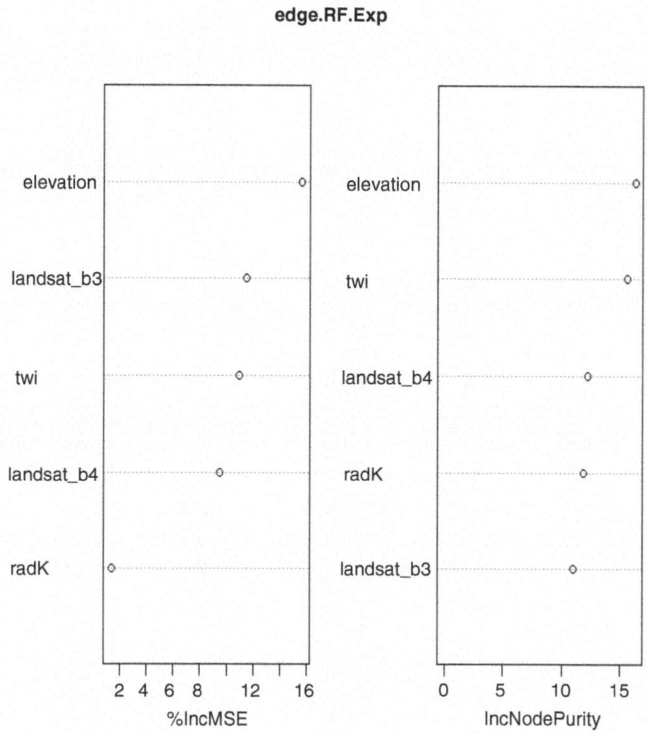

Fig. 5.9 Covariate importance (ranking of predictors) from Random Forest model fitting

There is a lot of talk about how the variable importance is measured in Random Forest models. So it is probably best to quote from the source:

> Here are the definitions of the variable importance measures. For each tree, the prediction accuracy on the out-of-bag portion of the data is recorded. Then the same is done after permuting each predictor variable. The difference between the two accuracy measurements are then averaged over all trees, and normalized by the standard error. For regression, the MSE is computed on the out-of-bag data for each tree, and then the same computed after permuting a variable. The differences are averaged and normalized by the standard error. If the standard error is equal to 0 for a variable, the division is not done (but the measure is almost always equal to 0 in that case). For the node purity, it is the total decrease in node impurities from splitting on the variable, averaged over all trees. For classification, the node impurity is measured by the Gini index. For regression, it is measured by residual sum of squares.

The important detail in the quotation above is the variable importance is measured based on "out-of-bag" samples. Which means observation not included in the model. Another detail is that they are based on an MSE accuracy measure; in this case the difference, when a covariate is included and excluded in a tree model. The value is averaged over all trees.

So lets validate the `edge.RF.Exp` model both internally and externally.

```
# Internal validation
RF.pred.C <- predict(edge.RF.Exp, newdata = DSM_data[training, ])
goof(observed = DSM_data$log_cStock0_5[training], predicted = RF.pred.C)

##          R2 concordance        MSE      RMSE         bias
## 1 0.877938   0.8466861 0.0686595 0.2620296 -0.004770852

# External validation
RF.pred.V <- predict(edge.RF.Exp, newdata = DSM_data[-training, ])
goof(observed = DSM_data$log_cStock0_5[-training], predicted = RF.pred.V)

##          R2 concordance        MSE      RMSE       bias
## 1 0.326027   0.4620323 0.1802985 0.4246157 -0.0841267
```

Essentially these results are quite similar to those of the validations from the other models. One needs to be careful about accepting very good model fit results without a proper external validation. This is particularly poignant for Random Forest models which have a tendency to provide excellent calibration results which can often give the wrong impression about its suitability for a given application. Therefore when using random forest models it pays to look at the validation on the out-of-bag samples when evaluating the goodness of fit of the model.

So lets look at the map resulting from applying the `edge.RF.Exp` models to the covariates (Fig. 5.10).

```
map.RF.r1 <- predict(covStack, edge.RF.Exp, "cStock_0_5_RF.tif",
format = "GTiff", datatype = "FLT4S", overwrite = TRUE)
plot(map.RF.r1,
main = "Random Forest model predicted 0-5cm log carbon stocks (0-5cm)")
```

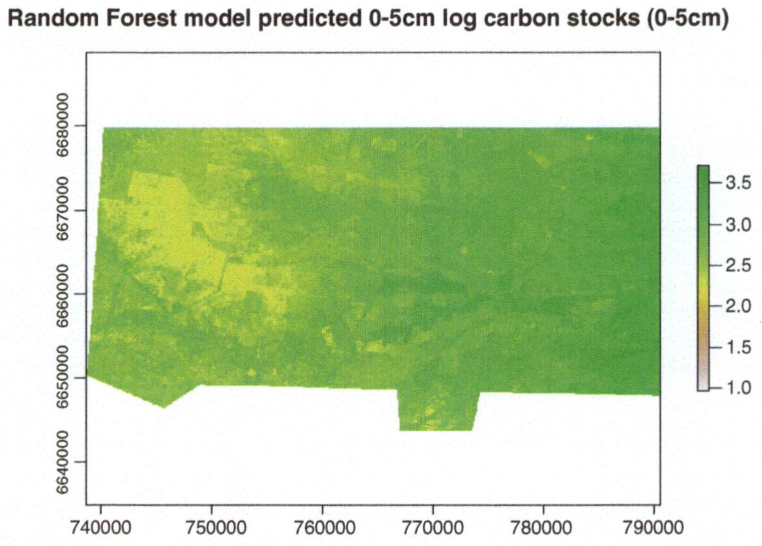

Fig. 5.10 Random Forest model predicted log SOC stock 0–5 cm Edgeroi

5.6 Advanced Work: Model Fitting with Caret Package

It becomes quickly apparent that there are many variants of prediction functions that could be used for DSM. As was observed, each of the models used have their relative advantages and disadvantages. Each also has their own specific parameterisations and quirks for fitting. Sometimes for the various parameters that are used for model training are chosen without any sort of optimisation, even due consideration sometimes. Sometimes we might be confronted with many possible model structures to use, it is often difficult to make a choice what to use, and just default with a model we know well or have used often without considering alternatives. This is where the `caret` R package http://topepo.github.io/caret/index.html comes into its own in terms of efficiency and streamlining the workflow for fitting models and optimising some of those parameter variables. As the dedicated website indicates (http://topepo.github.io/caret/index.html), the `caret` package (short for Classification And REgression Training) is a set of functions that attempt to streamline the process for creating predictive models. As we have seen, there are many different modeling functions in R. Some have different syntax for model training and/or prediction. The `caret` package provides a uniform interface to the various functions themselves, as well as a way to standardize common tasks (such parameter tuning and variable importance). There are currently over 300 model functions that the `caret` package interfaces with.

To begin, we first need to load the package into R:

```
library(caret)
```

The workhorse of the `caret` package is the `train` function. We can specify the model to be fitted in two ways:

```
# 1.
fit <- train(form = log_cStock0_5 ~ elevation + twi + radK +
landsat_b3 + landsat_b4, data = DSM_data, method = "lm")

# or 2.
fit <- train(x = DSM_data[, 4:8], y = DSM_data$log_cStock0_5,
method = "lm")
```

Using the `summary(fit)` command brings up the model parameter estimates, while the object `fit` also contains a summary of some useful model goodness of fit diagnostics such as the RMSE and R^2 statistics. You can control how model validation is done where options include simple goodness of fit calibration, k-fold cross-validation, and leave-one-out cross-validation. This option is controlled using the parameter `trControl` in the `train` function. The example below illustrates a 5-fold cross validation of a linear regression model with 10 repetitions.

```
fit <- train(x = DSM_data[, 4:8], y = DSM_data$log_cStock0_5, method = "lm",
trControl = trainControl(method = "repeatedcv", number = 5, repeats = 10))
```

```
fit
```

```
## Linear Regression
##
## 341 samples
##   5 predictor
##
## No pre-processing
## Resampling: Cross-Validated (5 fold, repeated 10 times)
## Summary of sample sizes: 273, 273, 272, 273, 273, 273, ...
## Resampling results
##
##   RMSE       Rsquared   RMSE SD     Rsquared SD
##   0.4918208  0.2197718  0.09759374  0.09535488
##
##
```

There are a lot of potential models that you could consider too for DSM. Check out http://topepo.github.io/caret/modelList.html or print them as below:

```
list_of_models <- modelLookup()
head(list_of_models)
```

```
##          model parameter          label forReg forClass probModel
## 1          ada      iter         #Trees  FALSE     TRUE      TRUE
## 2          ada  maxdepth Max Tree Depth  FALSE     TRUE      TRUE
## 3          ada        nu  Learning Rate  FALSE     TRUE      TRUE
## 4       AdaBag    mfinal         #Trees  FALSE     TRUE      TRUE
## 5       AdaBag  maxdepth Max Tree Depth  FALSE     TRUE      TRUE
## 6 AdaBoost.M1    mfinal         #Trees  FALSE     TRUE      TRUE
```

```
# The number of models caret interfaces with
nrow(list_of_models)
```

```
## [1] 372
```

You can choose which model to use in the `train` function with the `method` option. You will note that the fitting of the Cubist and Random Forest models below automatically attempt to optimise some of the fitting parameters, for example the `mtry` parameter for Random Forest. To look at what parameters can optimised for each model in caret we can use the `modelLookup` function.

```
# Cubist model
modelLookup(model = "cubist")
```

```
##    model  parameter       label forReg forClass probModel
## 1 cubist committees #Committees   TRUE    FALSE     FALSE
## 2 cubist  neighbors  #Instances   TRUE    FALSE     FALSE
```

```
fit_cubist <- train(x = DSM_data[, 4:8], y = DSM_data$log_cStock0_5,
method = "cubist", trControl = trainControl(method = "cv", number = 5))
```

```
# random forest model
modelLookup(model = "rf")

##   model parameter                         label forReg forClass probModel
## 1    rf      mtry #Randomly Selected Predictors   TRUE     TRUE      TRUE

fit_rf <- train(x = DSM_data[, 4:8], y = DSM_data$log_cStock0_5,
method = "rf", trControl = trainControl(method = "cv", number =5))
```

Using the fitted model, predictions can be achieved with the `predict` function:

```
# Cubist model
pred_cubist <- predict(fit_cubist, DSM_data)

# To raster data
pred_cubistMap <- predict(covStack, fit_cubist)
```

There is plenty of other added functionality of the `caret` package. In addition to the detailed resources mentioned above, it always pays to look over the help files that are associated with each function.

5.7 Regression Kriging

In the previous sections we looked at a few soil spatial prediction functions which at the most fundamental level, target the correlation between the target soil variable and the available covariate information. We fitted a number of models which included simple linear functions to non-linear functions such as regression trees to other more complicated data mining techniques (Cubist and Random Forest). In this section we will extend upon this DSM approach from what are called deterministic models to also include the spatially correlated residuals that result from fitting these models.

The approach we will now concentrate is a hybrid approach to modelling, whereby the predictions of the target variable are made via a deterministic method (regression model with covariate information) and a stochastic method where we determine the spatial auto-correlation of the model residuals with a variogram. The deterministic model essentially "detrends" the data, leaving behind the residuals for which we need to investigate whether there is additional spatial structure which could be added to the regression model predictions. These residuals are the random component of the $scorpan + e model$. This method is described as regression kriging and has formally been described in Odeh et al. (1995) and is synonymous with universal kriging (Hengl et al. 2007), which is the formal linear model procedure to this soil spatial modeling approach. The purpose of this exercise is to introduce some basic concepts of regression kriging. You will have already had some experience in regression models. We have also investigated briefly the fundamental concepts of kriging for which the variogram is central to.

5.7.1 *Universal Kriging*

The universal kriging function in R is found in the gstat package. It is useful from the view that both the regression model and variogram modeling of the residuals are handled together. Using universal kriging, one can efficiently derive prediction uncertainties by way of the kriging variance. A limitation of universal kriging in the true sense of the model parameter fitting is that the model is linear. The general preference is DSM studies is to used non-linear and recursive models that do not require strict model assumptions and assume a linear relationship between target variable and covariates.

One of the strict requirements of universal kriging in gstat is that the CRS (coordinate reference system) of the point data and covariates must be exactly the same. First we will take a subset of the data to use for an external validation. Unfortunately some of our data has to be sacrificed for this.

```
set.seed(123)
training <- sample(nrow(DSM_data), 0.7 * nrow(DSM_data))
cDat <- DSM_data[training, ]

coordinates(cDat) <- ~x + y
crs(cDat) <- "+proj=utm +zone=55 +south +ellps=WGS84 +datum=WGS84
+units=m +no_defs"
crs(covStack) = crs(cDat)

# check
crs(cDat)

## CRS arguments:
##  +proj=utm +zone=55 +south +ellps=WGS84 +datum=WGS84 +units=m
## +no_defs +towgs84=0,0,0

crs(covStack)

## CRS arguments:
##  +proj=utm +zone=55 +south +ellps=WGS84 +datum=WGS84 +units=m
## +no_defs +towgs84=0,0,0
```

Now lets parametise the universal kriging model, and we will use all of the available covariates.

```
library(gstat)
vgm1 <- variogram(log_cStock0_5 ~ elevation + twi + radK +
landsat_b3 + landsat_b4, cDat, width = 250, cressie = TRUE,
cutoff = 10000)
mod <- vgm(psill = var(cDat$log_cStock0_5), "Exp", range = 5000, nugget = 0)
model_1 <- fit.variogram(vgm1, mod)
model_1

##   model     psill    range
## 1   Nug 0.0000000   0.0000
```

```
## 2   Exp 0.1611552 233.7939

# Universal kriging model
gUK <- gstat(NULL, "log.carbon", log_cStock0_5 ~ elevation + twi + radK +
landsat_b3 + landsat_b4, cDat, model = model_1)
```

Using the validation data we can assess the performance of universal kriging using the `goof` function.

```
vDat <- DSM_data[-training, ]
coordinates(vDat) <- ~x + y
crs(vDat) <- "+proj=utm +zone=55 +south +ellps=WGS84 +datum=WGS84
+units=m +no_defs"
crs(vDat)

## CRS arguments:
##  +proj=utm +zone=55 +south +ellps=WGS84 +datum=WGS84 +units=m
## +no_defs +towgs84=0,0,0

# make the predictions
UK.preds.V <- as.data.frame(krige(log_cStock0_5 ~ elevation + twi
+ radK + landsat_b3 + landsat_b4, cDat, model = model_1,
newdata = vDat))

## [using universal kriging]

goof(observed = DSM_data$log_cStock0_5[-training],
predicted = UK.preds.V[,3])

##            R2 concordance       MSE      RMSE        bias
## 1 0.1404692   0.3713846 0.2813217 0.5303977 -0.08085698
```

The universal kriging model performs more-or-less the same as the MLR model that was fitted earlier.

Applying the universal kriging model spatially is facilitated through the `interpolate` function from `raster`. One can also use the `clusterR` function used earlier in order to speed things up a bit by applying the model over multiple compute nodes. Kriging results in two main outputs: the prediction and the prediction variance. When using the `interpolate` function we can control the output by changing the `index` parameter. The below script results in the maps on Fig. 5.11.

```
par(mfrow = c(1, 2))
# predictions
UK.P.map <- interpolate(covStack, gUK, xyOnly = FALSE, index = 1)
plot(UK.P.map, main = "Universal kriging predictions")

# prediction variance
UK.Pvar.map <- interpolate(covStack, gUK, xyOnly = FALSE, index = 2)
plot(UK.Pvar.map, main = "Universal krging prediction variance")
```

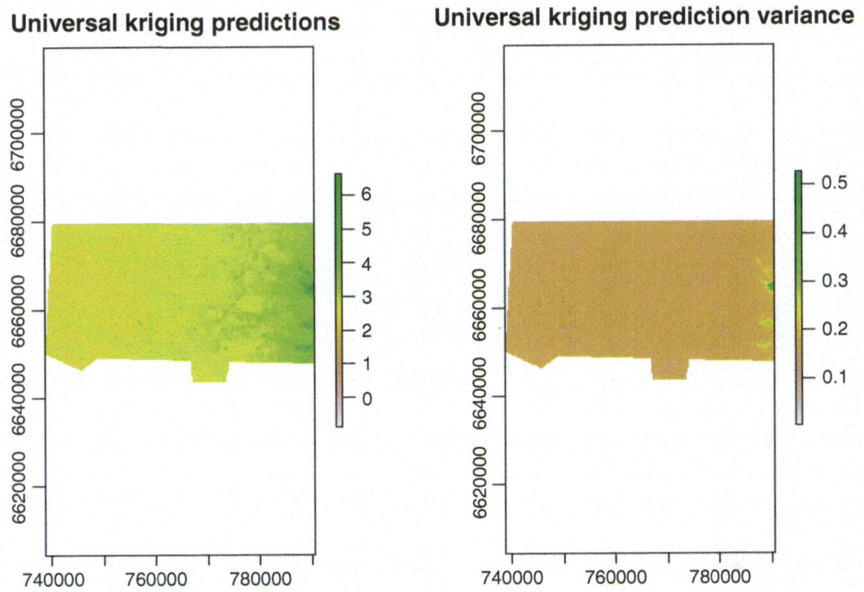

Fig. 5.11 Universal kriging prediction and prediction variance of log SOC stock 0–5 cm Edgeroi

5.7.2 *Regression Kriging with Cubist Models*

Even though we do not expect to achieve much by modeling the spatial structure of the model residuals using the Edgeroi data, this following example will provide the steps one would use to perform regression kriging that incorporates a complex model structure such as a data mining algorithm. Here we will use the Cubist model that was used earlier. Lets start from the beginning.

```
set.seed(123)
training <- sample(nrow(DSM_data), 0.7 * nrow(DSM_data))
mDat <- DSM_data[training, ]

# fit the model
edge.cub.Exp <- cubist(x = mDat[, c("elevation", "twi",  "radK",
"landsat_b3","landsat_b4")], y = mDat$log_cStock0_5, cubistControl
(unbiased = TRUE, rules = 100,extrapolation = 5, sample = 0,
label = "outcome"), committees = 1)
```

Now derive the model residual which is the model prediction subtracted from the residual.

```
mDat$residual <- mDat$log_cStock0_5 - predict(edge.cub.Exp,
newdata = mDat)
mean(mDat$residual)

## [1] 0.01197274
```

If you check the histogram of these residuals you will find that the mean is around zero and the data seems normally distributed. Now we can assess the residuals for any model structure,

```
coordinates(mDat) <- ~x + y
crs(mDat) <- "+proj=utm +zone=55 +south +ellps=WGS84 +datum=WGS84
+units=m +no_defs"

vgm1 <- variogram(residual ~ 1, mDat, width = 250, cressie = TRUE,
cutoff = 10000)
mod <- vgm(psill = var(mDat$residual), "Sph", range = 5000,
nugget = 0)
model_1 <- fit.variogram(vgm1, mod)
model_1

##   model      psill    range
## 1   Nug 0.02631769   0.0000
## 2   Sph 0.02921591 513.7697

# Residual kriging model
gRK <- gstat(NULL, "RKresidual", residual ~ 1, mDat,
model = model_1)
```

With the two model components together, we can now compare the external validation statistics of using the Cubist model only and with the Cubist model and residual variogram together.

```
# Cubist model only
Cubist.pred.V <- predict(edge.cub.Exp, newdata = DSM_data[-training, ])

# Cubist model with residual variogram
vDat <- DSM_data[-training, ]
coordinates(vDat) <- ~x + y
crs(vDat) <- "+proj=utm +zone=55 +south +ellps=WGS84 +datum=WGS84
+units=m +no_defs"

# make the residual predictions
RK.preds.V <- as.data.frame(krige(residual ~ 1, mDat, model = model_1,
newdata = vDat))

## [using ordinary kriging]

# Sum the two components together
RK.preds.fin <- Cubist.pred.V + RK.preds.V[, 3]

# validation cubist only
goof(observed = DSM_data$log_cStock0_5[-training],
predicted = Cubist.pred.V)

##          R2 concordance       MSE     RMSE       bias
## 1 0.1361656   0.3671604 0.3132954 0.559728 -0.1077685
```

```
# validation regression kriging with cubist model
goof(observed = DSM_data$log_cStock0_5[-training],
predicted = RK.preds.fin)

##          R2 concordance       MSE      RMSE       bias
## 1 0.1167233   0.3440912 0.3546901 0.5955586 -0.1001317
```

These results confirm that there was no advantage in performing regression kriging with this particular data. In any case, to apply the regression kriging model here, it requires three steps: First apply the Cubist model, then apply the residual kriging, then finally add both maps together. The script below illustrates how this is done, and the resulting maps are shown on Fig. 5.12.

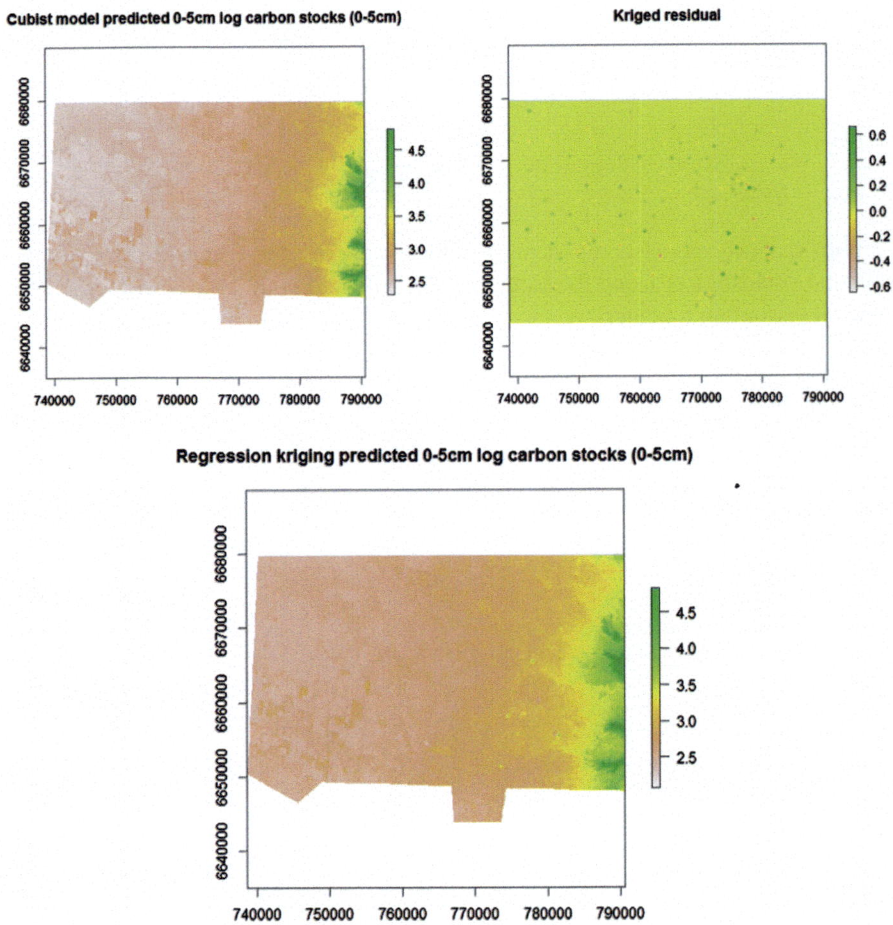

Fig. 5.12 Regression kriging predictions with cubist models. Log carbon stock (0–5 cm) Edgeroi

```
par(mfrow = c(3, 1))
map.RK1 <- predict(covStack, edge.cub.Exp,
filename = "cStock_0_5_cubistRK.tif",
format = "GTiff", datatype = "FLT4S", overwrite = TRUE)
plot(map.RK1, main = "Cubist model predicted 0-5cm log
carbon stocks (0-5cm)")

map.RK2 <- interpolate(covStack, gRK, xyOnly = TRUE, index = 1,
filename = "cStock_0_5_residualRK.tif", format = "GTiff",
datatype = "FLT4S", overwrite = TRUE)
plot(map.RK2, main = "Kriged residual")

pred.stack <- stack(map.RK1, map.RK2)
map.RK3 <- calc(pred.stack, fun = sum,
filename = "cStock_0_5_finalPredRK.tif",
format = "GTiff", progress = "text", overwrite = T)
plot(map.RK3,
main = "Regression kriging predicted 0-5cm log
carbon stocks (0-5cm)")
```

References

Breiman L (2001) Random forests. Mach Learn 41:5–32

Brus D, Kempen B, Heuvelink G (2011) Sampling for validation of digital soil maps. Eur J Soil Sci 62(3):394–407

de Gruijter J, Brus D, Bierkens M, Knotters M (2006) Sampling for natural resource monitoring. Springer, Berlin/Heidelberg

Grimm R, Behrens T, Marker M, Elsenbeer H (2008) Soil organic carbon concentrations and stocks on Barro Colorado Island: digital soil mapping using random forests analysis. Geoderma 146:102–113

Hastie T, Tibshirani R, Friedman J (2001) The elements of statistical learning. Springer, New York

Hengl T, Heuvelink GBM, Rossiter DG (2007) About regression kriging: from equations to case studies. Comput Geosci 33:1301–1315

Lin LI (1989) A concordance correlation coefficient to evaluate reproducibility. Biometrics 45:255–268

Odeh IOA, McBratney AB, Chittleborough DJ (1995) Further results on prediction of soil properties from terrain attributes: heterotopic co-kriging and regression kriging. Geoderma 67:215–226

Quinlan JR (1992) Learning with continuous classes. In: Proceedings of AI92, 5th Australian conference on artificial intelligence. World Scientific, Singapore, pp 343–348

Chapter 6
Categorical Soil Attribute Modeling and Mapping

The other form of soil spatial prediction functions are those dealing with categorical target variables such as soil classes. Naturally the models we will use further on are not specific to soil classes but can be generally applied to any type of categorical data too.

In the examples below we will demonstrate the prediction of soil-landscape classes termed: Terrons. Terron relates to a soil and landscape concept that has an associated model. The concept was first described by Carre and McBratney (2005). The embodiment of Terron is a continuous soil-landscape unit or class which combines soil knowledge, landscape information, and their interactions together. Malone et al. (2014) detailed an approach for defining Terrons in the Lower Hunter Valley, NSW, Australia. This area is a prominent wine-growing region of Australia, and the development of Terrons is a first step in the realization of viticultural terroir—an established concept of identity that incorporates much more than just soil and landscape qualities (Vaudour et al. 2015). In the work by Malone et al. (2014) they defined 12 Terrons for the Hunter Valley, which are distinguished by such soil and landscape characteristics as: geomorphometric attributes (derived from a digital elevation model) and specific attributes pertaining to soil pH, clay percentage, soil mineralogy (clay types and presence of iron oxides), continuous soil classes, and presence or absence of marl. The brief in this chapter is to predict the Terron classes across the prescribed Lower Hunter Valley, given a set of observations (sampled directly from the Terron map produced by Malone et al. (2014)).

By now you will be familiar with the process of fitting models and plotting/mapping the outputs. Thus in many ways, categorical data modeling is similar (in terms of implementation) with prediction models of continuous variables. The example in the previous chapter demonstrating the use of the `caret` package can also be similarly applied for categorial variables too, where you will find many model functions suited to that type of data with that package.

B.P. Malone et al., *Using R for Digital Soil Mapping*,
Progress in Soil Science, DOI 10.1007/978-3-319-44327-0_6

In this chapter we will have a look at a few different classification models:

1. Multinomial logistic regression
2. Data mining with the C5 algorithm
3. Random Forest

6.1 Model Validation of Categorical Prediction Models

The special characteristic of categorical data and its prediction within models, is that a class is either predicted or it is not. For binary variables. the prediction is either a *yes* or *no*, *black* or *white*, *present* or *absent* etc. For multinomial variables, there are more than 2 classes, for example either *black*, *grey*, or *white* etc (which could actually be an ordinal categorical classification, rather than nominal).

There is no in between; rather discrete entities. Exceptions are that some models do estimate the probability of the existence of a particular class, which will be touched on later. Additionally, there are methods of fuzzy classification which are common in the soil sciences (McBratney and Odeh 1997), but will not be covered in this section. Discrete categories and models for their prediction require other measures of validation than those that were used for continuous variables. The most important quality measures are described in Congalton (1991) and include:

1. Overall accuracy
2. User's accuracy
3. Producer's accuracy
4. Kappa coefficient of agreement

Using a contrived example, each of these quality measures will be illustrated. We will make a 4 × 4 matrix, and call it `con.mat`, and append some column and row names—in this case Australian Soil Classification Order codes. We then populate the matrix with some more-or-less random integers.

```
con.mat <- matrix(c(5, 0, 1, 2, 0, 15, 0, 5, 0, 1, 31, 0, 0,
10, 2, 11), nrow = 4, ncol = 4)
rownames(con.mat) <- c("DE", "VE", "CH", "KU")
colnames(con.mat) <- c("DE", "VE", "CH", "KU")
con.mat
```

```
##    DE VE CH KU
## DE  5  0  0  0
## VE  0 15  1 10
## CH  1  0 31  2
## KU  2  5  0 11
```

`con.mat` takes the form of a confusion matrix, and ones such as this are often the output of a classification model. If we summed each of the columns (using the `colSums` function), we would obtain the total number of observations for each soil class. Having column sums reflecting the number of observations is a widely used convention in classification studies.

```
colSums(con.mat)

## DE VE CH KU
##  8 20 32 23
```

Similarly, if we summed each of the rows we would retrieve the total number of predictions of each soil class. The predictions could have been made through any sort of model or classification process.

```
rowSums(con.mat)

## DE VE CH KU
##  5 26 34 18
```

Therefore, the numbers on the diagonal of the matrix will indicate fidelity between the observed class and the subsequent prediction. Numbers on the off-diagonals indicate a mis-classification or error. Overall accuracy is therefore computed by dividing the total correct (i.e., the sum of the diagonal) by the total number of observations (sum of the column sums).

```
ceiling(sum(diag(con.mat))/sum(colSums(con.mat)) * 100)

## [1] 75
```

Accuracy of individual classes can be computed in a similar manner. However, there is a choice of dividing the number of correct predictions for each class by either the totals (observations or predictions) in the corresponding columns or rows respectively. Traditionally, the total number of correct predictions of a class is divided by the total number of observations of that class (i.e. the column sum). This accuracy measure indicates the probability of an observation being correctly classified and is really a measure of omission error, or the "producer's accuracy". This is because the producer of the model is interested in how well a certain class can be predicted.

```
ceiling(diag(con.mat)/colSums(con.mat) * 100)

## DE VE CH KU
## 63 75 97 48
```

Alternatively, if the total number of correct predictions of a class is divided by the total number of predictions that were predicted in that category, then this result is a measure of commission error, or "user's accuracy". This measure is indicative of the probability that a prediction on the map actually represents that particular category on the ground or in the field.

```
ceiling(diag(con.mat)/rowSums(con.mat) * 100)

##  DE  VE  CH  KU
## 100  58  92  62
```

So if we use the DE category as an example, the "model" predicts this class correctly 63 % of the time, but when it is actually predicted it is correct 100 % of the time.

The Kappa coefficient is another statistical measure of the fidelity between observations and predictions of a classification. The calculation is based on the difference between how much agreement is actually present ("observed" agreement) compared to how much agreement would be expected to be present by chance alone ("expected" agreement). The observed agreement is simply the overall accuracy percentage. We may also want to know how different the observed agreement is from the expected agreement. The Kappa coefficient is a measure of this difference, standardized to lie on a -1 to 1 scale, where 1 is perfect agreement, 0 is exactly what would be expected by chance, and negative values indicate agreement less than chance, i.e., potential systematic disagreement between observations and predictions. The Kappa coefficient is defined as:

$$K = \frac{p_o - p_e}{1 - P_e} \tag{6.1}$$

where p_o is the overall or observed accuracy, and p_e is the expected accuracy, where:

$$p_e = \sum_{i=1}^{n} (\frac{colSum_i}{TO}) \times (\frac{rowSum_i}{TO}) \tag{6.2}$$

TO is the total number of observations and n is the number of classes. Rather than scripting the above equations, the kappa coefficient together with the other accuracy measures are contained in a function called goofcat in the ithir package. As we already have a confusion matrix prepared, we can enter it directly into the function as in the script below.

```
goofcat(conf.mat = con.mat, imp = TRUE)

## $confusion_matrix
##    DE VE CH KU
## DE  5  0  0  0
## VE  0 15  1 10
## CH  1  0 31  2
## KU  2  5  0 11
##
## $overall_accuracy
## [1] 75
##
```

```
## $producers_accuracy
## DE VE CH KU
## 63 75 97 48
##
## $users_accuracy
##  DE  VE  CH  KU
## 100  58  92  62
##
## $kappa
## [1] 0.6389062
```

A rule of thumb as indicated in Landis and Koch (1977) for the interpretation of the Kappa coefficient is:

< Less than chance agreement.
0.01–0.20 Slight agreement.
0.21–0.40 Fair agreement.
0.41–0.60 Moderate agreement.
0.61–0.80 Substantial agreement.
0.80–0.99 Almost perfect agreement.

We will be using these prediction quality indices for categorical variable prediction models in the following examples.

6.2 Multinomial Logistic Regression

Multinomial logistic regression is used to model nominal outcome variables, in which the log odds of the outcomes are modeled as a linear combination of the predictor variables. Because we are dealing with categorical variables, it is necessary that logistic regression take the natural logarithm of the odds (log-odds) to create a continuous criterion. The logit of success is then fit to the predictors using regression analysis. The results of the logit, however are not intuitive, so the logit is converted back to the odds via the inverse of the natural logarithm, namely the exponential function. Therefore, although the observed variables in logistic regression are categorical, the predicted scores are modeled as a continuous variable (the logit). The logit is referred to as the link function in logistic regression. As such, although the output in logistic regression will be multinomial, the logit is an underlying continuous criterion upon which linear regression is conducted. This means that for logistic regression we are able to return the most likely or probable prediction (class) as well as the probabilities of occurrence for all the other classes considered. Some discussion of the theoretical underpinnings of multinomial logistic regression, and importantly its application in DSM is given in Kempen et al. (2009).

In R we can use the multinom function from the nnet package to perform logistic regression. The are other implementations of this model in R, so it is worth a look to compare and contrast them. Fitting multinom is just like fitting a linear model as seen below.

As described earlier, the data to be used for the following modelling exercises are Terron classes as sampled from the map presented in Malone et al. (2014).The sample data contains 1000 entries of which there are 12 different Terron classes. Before getting into the modeling, we first load in the data and then perform the covariate layer intersection using the suite of environmental variables contained in the hunterCovariates data object in the ithir package. The small selection of covariates cover an area of approximately 220 km^2 at a spatial resolution of 25 m. They include those derived from a DEM: altitude above channel network (AACN), solar light insolation and terrain wetness index (TWI). Gamma radiometric data (total count) is also included together with a surface that depicts soil drainage in form of a continuous index (ranging from 0 to 5). These 5 different covariate layers are stacked together via a rasterStack.

```
library(ithir)
library(sp)
library(raster)

data(terron.dat)
data(hunterCovariates)
```

Transform the terron.dat data to a SpatialPointsDataFrame.

```
names(terron.dat)

## [1] "x"      "y"      "terron"

coordinates(terron.dat) <- ~x + y
```

As these data are of the same spatial projection as the hunterCovariates, there is no need to perform a coordinate transformation. So we can perform the intersection immediately.

```
DSM_data <- extract(covStack, terron.dat, sp = 1,
method = "simple")
DSM_data <- as.data.frame(DSM_data)
str(DSM_data)

## 'data.frame': 1000 obs. of  8 variables:
##  $ x                 : num  346535 334760 340910
                          336460 344510 ...
##  $ y                 : num  6371941 6375841 6377691
                          6382041 6378116 ...
##  $ terron            : Factor w/ 12 levels "1","2",
                          "3","4",..: 3 4 ...
##  $ AACN              : num  37.544 25.564 32.865
                          0.605 9.516 ...
```

```
## $ Drainage.Index    : num  4.72 4.78 2 4.19 4.68 ...
## $ Light.Insolation : num  1690 1736 1712 1712 1677 ...
## $ TWI               : num  11.5 13.8 13.4 18.6 19.8 ...
## $ Gamma.Total.Count: num  380 407 384 388 454 ...
```

It is always good practice to check to see if any of the observational data returned any NA values for any one of the covariates. If there is NA values, it indicates that the observational data is outside the extent of the covariate layers. It is best to remove these observations from the data set.

```
which(!complete.cases(DSM_data))

## integer(0)

DSM_data <- DSM_data[complete.cases(DSM_data), ]
```

Now for model fitting. The target variable is `terron`. So lets just use all the available covariates in the model. We will also subset the data for an external validation i.e. random hold back validation.

```
library(nnet)

set.seed(655)
training <- sample(nrow(DSM_data), 0.7 * nrow(DSM_data))
hv.MNLR <- multinom(terron ~ AACN + Drainage.Index
+ Light.Insolation + TWI + Gamma.Total.Count,
data = DSM_data[training, ])
```

Using the `summary` function allows us to see the linear models for each Terron class, which are the result of the log-odds of each soil class modeled as a linear combination of the covariates. We can also see the probabilities of occurrence for each Terron class at each observation location by using the `fitted` function.

```
summary(hv.MNLR)

# Estimate class probabilities
probs.hv.MNLR <- fitted(hv.MNLR)

# return top of data frame of probabilites
head(probs.hv.MNLR)
```

Subsequently, we can also determine the most probable Terron class using the the `predict` function.

```
pred.hv.MNLR <- predict(hv.MNLR)
summary(pred.hv.MNLR)

##   1   2   3   4   5   6   7   8   9  10  11  12
##  21   7  60  62 110  73 169 115  18  29  12  24
```

Lets now perform an internal validation of the model to assess its general performance. Here we use the `goofcat` function, but this time we import the two vectors into the function which correspond to the observations and predictions respectively.

```
goofcat(observed = DSM_data$terron[training],
predicted = pred.hv.MNLR)

## $confusion_matrix
##      1 2  3  4  5  6  7  8  9 10 11 12
## 1   13 3  1  2  0  0  1  0  1  0  0  0
## 2    1 5  0  0  0  0  0  0  1  0  0  0
## 3    0 1 35  2  0  0  8  0  2  3  6  3
## 4    4 0  9 30  1  1  3  3  5  0  6  0
## 5    0 0  0  0 56 10  1 10 13 17  1  2
## 6    0 0  0  1 16 47  0  6  1  2  0  0
## 7    0 0 14  7  4  0 92 16  7  8  9 12
## 8    0 0  0  4 10  7 14 57  4  6 13  0
## 9    0 0  0  5  3  0  2  2  5  0  1  0
## 10   0 0  0  0  4  1  2  6  3 13  0  0
## 11   0 0  4  3  0  0  0  3  0  0  2  0
## 12   1 0  2  1  0  0  1  0  0  4  0 15
##
## $overall_accuracy
## [1] 53
##
## $producers_accuracy
##  1  2  3  4  5  6  7  8  9 10 11 12
## 69 56 54 55 60 72 75 56 12 25  6 47
##
## $users_accuracy
##  1  2  3  4  5  6  7  8  9 10 11 12
## 62 72 59 49 51 65 55 50 28 45 17 63
##
## $kappa
## [1] 0.4637285
```

Similarly, performing the external validation requires first using the `pred.hv.MNLR` model to predict on the withheld points.

```
V.pred.hv.MNLR <- predict(hv.MNLR,
newdata = DSM_data[-training, ])
goofcat(observed = DSM_data$terron[-training],
predicted = V.pred.hv.MNLR)

## $confusion_matrix
##     1 2  3 4  5  6  7  8 9 10 11 12
## 1   1 1  1 2  0  0  0  0 1  0  0  0
## 2   5 0  2 0  0  0  0  0 1  1  0  0
## 3   0 0 13 1  0  0  5  0 1  2  2  2
## 4   2 3  9 8  0  0  0  4 8  0  3  0
```

```
## 5   0 0   0 0 21  7  0  8 6  3  0  2
## 6   0 0   0 1  8 18  0  5 1  0  0  0
## 7   0 0   8 4  0  0 38  9 4  5  2  5
## 8   0 0   0 3  3  0  6 15 1  3  1  0
## 9   0 0   0 1  2  0  0  1 0  0  0  0
## 10  0 0   0 0  4  1  1  4 1  9  2  0
## 11  0 0   0 0  0  0  0  1 0  0  0  0
## 12  0 0   1 0  0  0  1  0 0  2  0  4
##
## $overall_accuracy
## [1] 43
##
## $producers_accuracy
##   1  2  3  4  5  6  7  8  9 10 11 12
## 13  0 39 40 56 70 75 32  0 36  0 31
##
## $users_accuracy
##   1  2  3  4  5  6  7  8  9 10 11 12
## 17  0 50 22 45 55 51 47  0 41  0 50
##
## $kappa
## [1] 0.3476539
```

Using the raster `predict` function is the method for applying the `hv.MNLR` model across the whole area. Note that the `clusterR` function can also be used here too if there is a requirement to perform the spatial prediction across multiple compute nodes. Note also that it is also possible for multinomial logistic regression to create the map of the most probable class, as well as the probabilities for all classes. The first script example below is for mapping the most probable class which is specified by setting the `type` parameter to "`class`". If probabilities are required "`probs`" would be used for the `type` parameter, together with specifying an `index` integer to indicate which class probabilities you wish to map. The second script example below shows the parametisation for predicting the probabilities for Terron class 1.

```
# class prediction
map.MNLR.c <- predict(covStack, hv.MNLR, type = "class",
filename = "hv_MNLR_class.tif",format = "GTiff",
overwrite = T, datatype = "INT2S")

# class probabilities
map.MNLR.p <- predict(covStack, hv.MNLR, type = "probs",
index = 1, filename = "edge_MNLR_probs1.tif", format = "GTiff",
overwrite = T, datatype = "FLT4S")
```

Plotting the resulting class map is not as straightforward as for mapping continuous variables. A solution is scripted below which uses an associated package to `raster` called `rasterVis`. You will also note the use of explicit colors for each Terron class as they were the same colors used in the Terron map presented

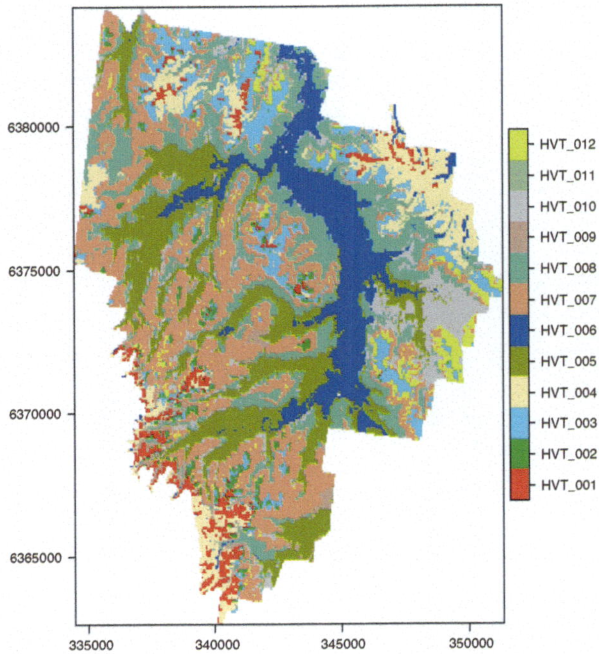

Fig. 6.1 Hunter Valley most probable Terron class map created using multinomial logistic regression model

by Malone et al. (2014). The colors are defined in terms of HEX color codes. A very good resource for selecting colors or deciding on color ramps for maps is *colorbrewer* located at http://colorbrewer2.org/. The script below produces the map in Fig. 6.1.

```
map.MNLR.c <- as.factor(map.MNLR.c)

## Add a land class column to the Raster Attribute Table
rat <- levels(map.MNLR.c)[[1]]
rat[["terron"]] <- c("HVT_001", "HVT_002", "HVT_003", "HVT_004",
    "HVT_005", "HVT_006", "HVT_007", "HVT_008","HVT_009",
    "HVT_010", "HVT_011", "HVT_012")
levels(map.MNLR.c) <- rat

## HEX colors
area_colors <- c("#FF0000", "#38A800", "#73DFFF", "#FFEBAF",
    "#A8A800", "#0070FF", "#FFA77F", "#7AF5CA", "#D7B09E",
    "#CCCCCC", "#B4D79E", "#FFFF00")
# plot
levelplot(map.MNLR.c, col.regions = area_colors,
xlab = "", ylab = "")
```

6.3 C5 Decision Trees

The C5 decision tree model is available in the `C50` package. The function `C5.0` fits classification tree models or rule-based models using Quinlans's C5.0 algorithm (Quinlan 1993).

Essentially we will go through the same process as we did for the multinomial logistic regression. The `C5.0` function and its internal parameters are similar in nature to that of the `Cubist` function for predicting continuous variables. The `trials` parameter lets you implement a "boosted" classification tree process, with the results aggregated at the end. There are also many other useful model tuning parameters in the `C5.0Control` parameter set too that are worth a look. Just see the help files for more information.

Using the same training and validation sets as before, we will fit the C5 model as in the script below.

```
library(C50)
hv.C5 <- C5.0(x = DSM_data[training, c("AACN", "Drainage.Index",
"Light.Insolation", "TWI", "Gamma.Total.Count")],
y = DSM_data$terron[training], trials = 1, rules = FALSE))
```

By calling the `summary` function, some useful model fit diagnostics are given. These include the tree structure, together with the covariates used and misclassification error of the model, as well as a rudimentary confusion matrix. A useful feature of the C5 model is that it can omit in an automated fashion, unnecessary covariates.

The `predict` function can either return the predicted class or the confidence of the predictions (which is controlled using the `type="prob"` parameter). The probabilities are quantified such that if an observation is classified by a single leaf of a decision tree, the confidence value is the proportion of training cases at that leaf that belong to the predicted class. If more than one leaf is involved (i.e., one or more of the attributes on the path has missing values), the value is a weighted sum of the individual leaves' confidences. For rule-classifiers, each applicable rule votes for a class with the voting weight being the rule's confidence value. If the sum of the votes for class C is W(C), then the predicted class P is chosen so that W(P) is maximal, and the confidence is the greater of (1), the voting weight of the most specific applicable rule for predicted class P; or (2) the average vote for class P (so, W(P) divided by the number of applicable rules for class P).Boosted classifiers are similar, but individual classifiers vote for a class with weight equal to their confidence value. Overall, the confidence associated with each class for every observation are made to sum to 1.

```
# return the class predictions
predict(hv.C5, newdata = DSM_data[training, ])

# return the class probabilities
predict(hv.C5, newdata = DSM_data[training, ], type = "prob")
```

So lets look at the calibration and validation statistics. First, the calibration statistics:

```
C.pred.hv.C5 <- predict(hv.C5, newdata = DSM_data[training, ])
goofcat(observed = DSM_data$terron[training],
predicted = C.pred.hv.C5)
```

```
## $confusion_matrix
##      1 2  3  4  5  6  7  8  9 10 11 12
## 1   17 9  4  4  0  0  0  0  1  0  0  1
## 2    0 0  0  0  0  0  0  0  0  0  0  0
## 3    0 0 42  4  0  0 13  0  4  3  7 14
## 4    2 0 12 37  0  1  5  6  9  1  9  1
## 5    0 0  0  0 48  2  1  9 14 18  0  0
## 6    0 0  0  0 22 55  0  3  2  1  1  0
## 7    0 0  7  6  7  0 95 18  9 12  5 14
## 8    0 0  0  4 12  8  2 61  3  4  9  2
## 9    0 0  0  0  0  0  0  0  0  0  0  0
## 10   0 0  0  0  5  0  8  6  0 14  7  0
## 11   0 0  0  0  0  0  0  0  0  0  0  0
## 12   0 0  0  0  0  0  0  0  0  0  0  0
##
## $overall_accuracy
## [1] 53
##
## $producers_accuracy
##  1  2  3  4  5  6  7  8  9 10 11 12
## 90  0 65 68 52 84 77 60  0 27  0  0
##
## $users_accuracy
##   1   2   3   4   5   6   7   8   9  10  11  12
##  48 NaN  49  45  53  66  55  59 NaN  35 NaN NaN
##
## $kappa
## [1] 0.4618099
```

It will be noticed that some of the Terron classes failed to be predicted by the fitted model. For example Terron classes 2, 9, 11, and 12 were all predicted as being a different class. All observations of Terron class 2 were predicted as Terron class 1. Doing the external validation we return the following:

```
V.pred.hv.C5 <- predict(hv.C5, newdata = DSM_data[-training, ])
goofcat(observed = DSM_data$terron[-training],
predicted = V.pred.hv.C5)
```

```
## $confusion_matrix
##     1 2  3 4  5  6  7  8 9 10 11 12
## 1   7 3  5 1  0  0  1  0 2  1  0  1
## 2   0 0  0 0  0  0  0  0 0  0  0  0
## 3   0 0 14 1  0  0  4  0 2  2  2  6
## 4   1 1  5 9  0  0  4  3 7  0  0  0
```

```
## 5   0 0   0 0 20  2   0   7 5  2  0  1
## 6   0 0   0 0 10 22   0   7 0  1  0  0
## 7   0 0 10 5   0  0  35 11 2  7  5  3
## 8   0 0   0 3   6  2   4 16 2  5  2  2
## 9   0 0   0 0   0  0   0  0 0  0  0  0
## 10 0 0   0 1   2  0   3  3 4  7  1  0
## 11 0 0   0 0   0  0   0  0 0  0  0  0
## 12 0 0   0 0   0  0   0  0 0  0  0  0
##
## $overall_accuracy
## [1] 44
##
## $producers_accuracy
##  1  2  3  4  5  6  7  8  9 10 11 12
## 88  0 42 45 53 85 69 35  0 29  0  0
##
## $users_accuracy
##    1   2   3   4   5   6   7   8   9  10  11  12
##   34 NaN  46  30  55  56  45  39 NaN  34 NaN NaN
##
## $kappa
## [1] 0.3565075
```

And finally, creating the map derived from the `hv.C5` model using the raster `predict` function (Fig. 6.2). Note that the C5 model returned 0 % producer's

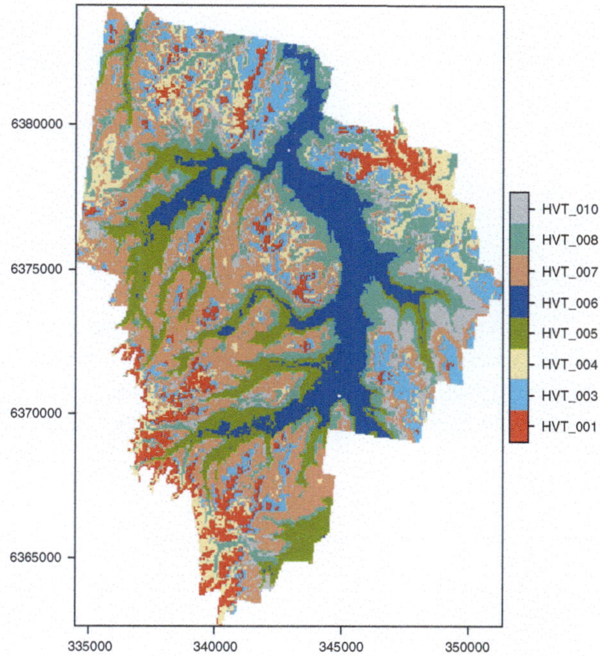

Fig. 6.2 Hunter Valley Terron class map created using C5 decision tree model

accuracy for Terron classes 2, 9, 11 and 12. These data account for only a small proportion of the data set, and/or, they may be similar to other existing Terron classes (based on the available predictive covariates). Consequently, they did not feature in the `hv.C5` model and ultimately, the final map.

```
# class prediction
map.C5.c <- predict(covStack, hv.C5, type = "class",
filename = "hv_C5_class.tif",format = "GTiff",
overwrite = T, datatype = "INT2S")

# plot
levelplot(map.C5.c, col.regions = area_colors,
xlab = "", ylab = "")
```

6.4 Random Forests

The final model we will look at is the Random Forest, which we should be familiar with now as this model type was examined during the continuous variable prediction methods section. It can also be used for categorical variables. Some useful extractor functions like `print` and `importance` give some useful information about the model performance.

```
library(randomForest)

hv.RF <- randomForest(terron ~ AACN + Drainage.Index
+ Light.Insolation + TWI + Gamma.Total.Count,
data = DSM_data[training, ], ntree = 500, mtry = 5)


# Output random forest model diognostics
print(hv.RF)

# output relative importance of each covariate
importance(hv.RF)
```

Three types of prediction outputs can be generated from Random Forest models, and are specified via the `type` parameter of the `predict` extractor functions. The different "types" are the `response` (predicted class), `prob` (class probabilities) or `vote` (vote count, which really just appears to return the probabilities).

```
# Prediction of classes
predict(hv.RF, type = "response", newdata = DSM_data[training, ])
```

```
# Class probabilities
predict(hv.RF, type = "prob", newdata = DSM_data[training, ])
```

From the diagnostics output of the `hv.C5` model the confusion matrix is automatically generated, except it was a different orientation to what we have been looking for previous examples. This confusion matrix was performed on what is called the OOB or out-of-bag data i.e. it validates the model/s dynamically with observations withheld from the model fit. So lets just evaluate the model as we have done for the previous models. For calibration:

```
C.pred.hv.RF <- predict(hv.RF, newdata = DSM_data[training, ])
goofcat(observed = DSM_data$terron[training],
predicted = C.pred.hv.RF)

## $confusion_matrix
##      1  2  3  4  5  6   7   8  9 10 11 12
## 1   19  0  0  0  0  0   0   0  0  0  0  0
## 2    0  9  0  0  0  0   0   0  0  0  0  0
## 3    0  0 65  0  0  0   0   0  0  0  0  0
## 4    0  0  0 55  0  0   0   0  0  0  0  0
## 5    0  0  0  0 94  0   0   0  0  0  0  0
## 6    0  0  0  0  0 66   0   0  0  0  0  0
## 7    0  0  0  0  0  0 124   0  0  0  0  0
## 8    0  0  0  0  0  0   0 103  0  0  0  0
## 9    0  0  0  0  0  0   0   0 42  0  0  0
## 10   0  0  0  0  0  0   0   0  0 53  0  0
## 11   0  0  0  0  0  0   0   0  0  0 38  0
## 12   0  0  0  0  0  0   0   0  0  0  0 32
##
## $overall_accuracy
## [1] 100
##
## $producers_accuracy
##   1   2   3   4   5   6   7   8   9  10  11  12
## 100 100 100 100 100 100 100 100 100 100 100 100
##
## $users_accuracy
##   1   2   3   4   5   6   7   8   9  10  11  12
## 100 100 100 100 100 100 100 100 100 100 100 100
##
## $kappa
## [1] 1
```

It seems quite incredible that this particular model is indicating a 100 % accuracy. Here it pays to look at the out-of-bag error of the `hv.RF` model for a better indication of the model goodness of fit. FOr the random holdback validation:

```
V.pred.hv.RF <- predict(hv.RF, newdata = DSM_data[-training, ])
goofcat(observed = DSM_data$terron[-training],
predicted = V.pred.hv.RF)
```

```
## $confusion_matrix
##     1 2  3  4  5  6  7  8 9 10 11 12
## 1   2 2  1  2  0  0  0  0 1  0  0  0
## 2   4 0  2  0  0  0  0  0 1  1  0  0
## 3   0 0 11  1  0  0  7  0 2  0  2  1
## 4   2 2  4 10  0  0  1  3 6  0  0  0
## 5   0 0  0  0 29  8  2  7 5  2  0  1
## 6   0 0  0  0  2 18  0  6 0  1  0  0
## 7   0 0  8  5  0  0 31  8 3  4  3  5
## 8   0 0  0  0  4  0  8 19 3  1  2  1
## 9   0 0  1  0  0  0  0  0 2  0  0  0
## 10  0 0  2  0  2  0  1  3 1 12  1  0
## 11  0 0  5  2  1  0  0  1 0  1  2  0
## 12  0 0  0  0  0  0  1  0 0  3  0  5
##
## $overall_accuracy
## [1] 47
##
## $producers_accuracy
##  1  2  3  4  5  6  7  8  9 10 11 12
## 25  0 33 50 77 70 61 41  9 48 20 39
##
## $users_accuracy
##  1  2  3  4  5  6  7  8  9 10 11 12
## 25  0 46 36 54 67 47 50 67 55 17 56
##
## $kappa
## [1] 0.4015957
```

So based on the model validation, the Random Forest performs quite similarly to the other models that were used before, despite a *perfect* performance based on the diagnostics of the calibration model.

And finally the map that results from applying the `hv.C5` model to the covariate rasters in shown on Fig. 6.3.

```
# class prediction
map.RF.c <- predict(covStack, hv.RF, filename = "hv_RF_class.tif",
format = "GTiff",overwrite = T, datatype = "INT2S")

# plot
levelplot(map.RF.c, col.regions = area_colors,
xlab = "", ylab = "")
```

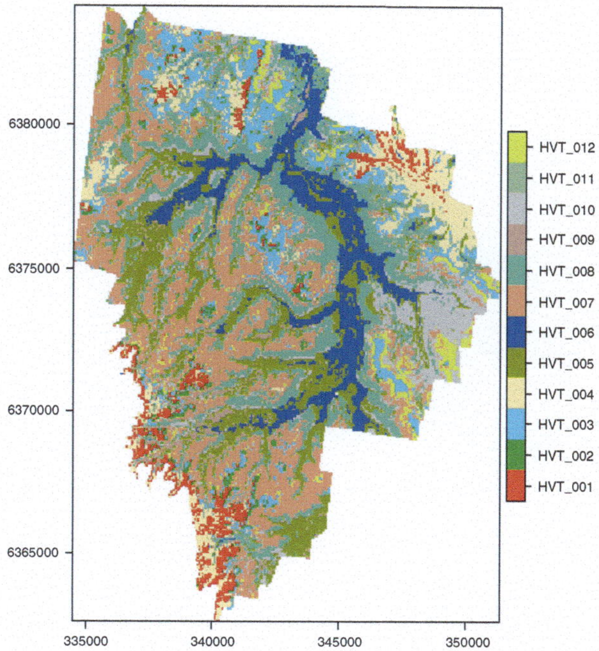

Fig. 6.3 Hunter Valley Terron class map created using random forest model

References

Carre F, McBratney AB (2005) Digital Terron mapping. Geoderma 128:340–353

Congalton RG (1991) A review of assessing the accuracy of classifications of remotely sensed data. Remote Sens Environ 37:35–46

Kempen B, Brus DJ, Heuvelink GBM, Stoorvogel JJ (2009) Updating the 1:50,000 Dutch soil map using legacy soil data: a multinomial logistic regression approach. Geoderma 151:311–326

Landis R, Koch GG (1977) The measurement of observer agreement for categorical data. Biometrics 33:159–174

Malone BP, Hughes P, McBratney AB, Minsany B (2014) A model for the identification of Terrons in the lower hunter valley, Australia. Geoderma Reg 1:31–47

McBratney AB, Odeh IOA (1997) Application of fuzzy sets in soil science: fuzzy logic, fuzzy measurement and fuzzy decisions. Geoderma 77:85–113

Quinlan JR (1993) C4.5: programs for machine learning. Morgan Kaufmann, San Mateo

Vaudour E, Costantini E, Jones GV, Mocali S (2015) An overview of the recent approaches to terroir functional modelling, footprinting and zoning. SOIL 1:287–312

Chapter 7
Some Methods for the Quantification of Prediction Uncertainties for Digital Soil Mapping

Soil scientists are quite aware of the current issues concerning the natural environment because our expertise is intimately aligned with their understanding and alleviation. We know that sustainable soil management alleviates soil degradation, improves soil quality and will ultimately ensure food security. Critical to better soil management is information detailing the soil resource, its processes and its variation across landscapes. Consequently, under the broad umbrella of "environmental monitoring", there has been a growing need to acquire quantitative soil information (Grimm and Behrens 2010; McBratney et al. 2003). The concerns of soil-related issues in reference to environmental management were raised by McBratney (1992) when stating that it is our duty as soil scientists, to ensure that the information we provide to the users of soil information is both accurate and precise, or at least of known accuracy and precision.

However, a difficulty we face is that soil can vary, seemingly erratically in the context of space and time (Webster 2000). Thus the conundrum in model-based predictions of soil phenomena is that models are not "error free". The unpredictability of soil variation combined with simplistic representations of complex soil processes inevitably leads to errors in model outputs.

We do not know the true character and processes of soils and our models are merely abstractions of these real processes. We know this; or in other words, in the absence of such confidence, we know we are uncertain about the true properties and processes that characterize soils (Brown and Heuvelink 2005). The key is therefore to determine to what extent our uncertainties are propagated through a model of which effect the final predictions of a real-world process.

In modeling exercises, uncertainty of the model output is the summation of the three main sources generally described as: model structure uncertainty, model parameter uncertainty and model input uncertainty (Brown and Heuvelink 2005; Minasny and McBratney 2002b). A detailed analysis of the contribution of each of the different sources of uncertainty is generally recommended. In this book chapter we will cover few approaches to estimate the uncertainty of model outputs. Essen-

B.P. Malone et al., *Using R for Digital Soil Mapping*,
Progress in Soil Science, DOI 10.1007/978-3-319-44327-0_7

tially what this means is that given a defined level of confidence, model predictions from digital soil mapping will be co-associated with the requisite prediction interval or range. The approaches for quantifying the prediction uncertainties are:

- Universal kriging prediction variance.
- Bootstrapping
- Empirical uncertainty quantification through data partitioning and cross validation.
- Empirical uncertainty quantification through fuzzy clustering and cross validation

The data that will be used in this chapter is a small data set of subsoil pH that has been collected since 2001 to present from the Lower Hunter Valley in New South Wales, Australia. The soil data covers an area of approximately 220 km^2. Validation of the quantification of uncertainty will be performed using a subset of these data. The mapping of the uncertainties will be conducted for a small region of the study area. The data for this section can be retrieved from the `ithir` package. The soil data is called `HV_subsoilpH` while the grids of environmental covariates is called `hunterCovariates_sub`.

7.1 Universal Kriging Prediction Variance

In the chapter regarding digital soil mapping of continuous variables, universal kriging was explored. This model is ideal from the perspective that both the correlated variables and model residuals are handled simultaneously. This model also automatically generates prediction uncertainty via the kriging variance. It is with this variance estimate that we can define a prediction interval. For this example and the following, a 90 % prediction interval will be defined for the mapping purposes. Although for validation, a number of levels of confidence will be defined and subsequently validated in order to assess the performance and sensitivity of the uncertainty estimates.

7.1.1 Defining the Model Parameters

First we need to load in all the libraries that are necessary for this section and load in the necessary data.

```
library(ithir)
library(sp)
library(rgdal)
library(raster)
library(gstat)
```

```
# Point data
data(HV_subsoilpH)
str(HV_subsoilpH)

## 'data.frame': 506 obs. of  14 variables:
##  $ X                        : num  340386 340345 340559
                                 340483 ...
##  $ Y                        : num  6368690 6368491 6369168 ...
##  $ pH60_100cm               : num  4.47 5.42 6.26 8.03 8.86 ...
##  $ Terrain_Ruggedness_Index: num  1.34 1.42 1.64 1.04 1.27 ...
##  $ AACN                     : num  1.619 0.281 2.301
                                 1.74 3.114 ...
##  $ Landsat_Band1            : int  57 47 59 52 62 53 47
                                 52 53 63 ...
##  $ Elevation                : num  103.1 103.7 99.9
                                 101.9 99.8 ...
##  $ Hillshading              : num  1.849 1.428 0.934
                                 1.517 1.652 ...
##  $ Light_insolation         : num  1689 1701 1722
                                 1688 1735 ...
##  $ Mid_Slope_Positon        : num  0.876 0.914 0.844
                                 0.848 0.833 ...
##  $ MRVBF                    : num  3.85 3.31 3.66 3.92 3.89 ...
##  $ NDVI                     : num  -0.143 -0.386 -0.197 ...
##  $ TWI                      : num  17.5 18.2 18.8 18 17.8 ...
##  $ Slope                    : num  1.79 1.42 1.01 1.49 1.83 ...

# Raster data
data(hunterCovariates_sub)
hunterCovariates_sub

## class       : RasterStack
## dimensions  : 249, 210, 52290, 11  (nrow, ncol, ncell, nlayers)
## resolution  : 25, 25  (x, y)
## extent      : 338422.3, 343672.3, 6364203, 6370428
## (xmin, xmax, ymin, ymax)
## coord. ref. : +proj=utm +zone=56 +south +ellps=WGS84
## +datum=WGS84 +units=m +no_defs
## names       : Terrain_Ruggedness_Index,       AACN,
## Landsat_Band1, Elevation, Hillshading, Light_insolation,
## Mid_Slope_Positon, MRVBF,        NDVI,           TWI,       Slope
## min values  :                   0.194371,     0.000000,
## 26.000000, 72.217499,    0.000677,       1236.662840,
## 0.000009,  0.000002,   -0.573034,     8.224325,     0.001708
## max values  :                  15.945321,  106.665482,
## 140.000000, 212.632507,   32.440960,    1934.199950, 0.956529,
## 4.581594,    0.466667,   20.393652,   21.809752
```

You will notice for `HV_subsoilpH` that these data have already been intersected with a number of covariates. The `hunterCovariates_sub` are a `rasterStack` of the same covariates (although the spatial extent is smaller).

Now to prepare the data for the universal kriging model.

```
# subset data for modeling
set.seed(123)
training <- sample(nrow(HV_subsoilpH), 0.7 * nrow(HV_subsoilpH))
cDat <- HV_subsoilpH[training, ]
vDat <- HV_subsoilpH[-training, ]
nrow(cDat)

## [1] 354

nrow(vDat)

## [1] 152
```

The `cDat` and `vDat` objects correspond to the model calibration and validation data sets respectively.

Now to prepare the data for the model

```
# coordinates
coordinates(cDat) <- ~X + Y

# remove CRS from grids
crs(hunterCovariates_sub) <- NULL
```

We will firstly use a step wise regression to determine a parsimonious model are the most important covariates.

```
# Full model
lm1 <- lm(pH60_100cm ~ Terrain_Ruggedness_Index + AACN
+ Landsat_Band1 + Elevation + Hillshading + Light_insolation
+ Mid_Slope_Positon + MRVBF + NDVI + TWI + Slope, data = cDat)

# Parsimous model
lm2 <- step(lm1, direction = "both", trace = 0)
as.formula(lm2)

## pH60_100cm ~ AACN + Landsat_Band1 + Hillshading
##   + Mid_Slope_Positon + MRVBF + NDVI + TWI

summary(lm2)

##
## Call:
## lm(formula = pH60_100cm ~ AACN + Landsat_Band1 + Hillshading +
##     Mid_Slope_Positon + MRVBF + NDVI + TWI, data = cDat)
##
## Residuals:
##     Min      1Q  Median      3Q     Max
## -2.9409 -0.8467 -0.1431  0.6870  3.2195
##
## Coefficients:
```

```
##                      Estimate Std. Error t value Pr(>|t|)
## (Intercept)           6.55363    1.00729   6.506 2.70e-10 ***
## AACN                  0.02652    0.00579   4.580 6.48e-06 ***
## Landsat_Band1        -0.04391    0.01119  -3.925 0.000104 ***
## Hillshading           0.07651    0.02139   3.576 0.000398 ***
## Mid_Slope_Positon     0.88822    0.31849   2.789 0.005582 **
## MRVBF                 0.28889    0.09624   3.002 0.002878 **
## NDVI                  5.88079    1.06282   5.533 6.22e-08 ***
## TWI                   0.11132    0.05657   1.968 0.049889 *
## ---
## Signif. codes:  0 '***' 0.001 '**' 0.01 '*' 0.05 '.' 0.1 ' ' 1
##
## Residual standard error: 1.185 on 346 degrees of freedom
## Multiple R-squared:  0.2447,Adjusted R-squared:  0.2294
## F-statistic: 16.01 on 7 and 346 DF,  p-value: < 2.2e-16
```

Now we can construct the universal kriging model using the step wise selected covariates.

```
vgm1 <- variogram(pH60_100cm ~ AACN + Landsat_Band1 +
Hillshading + Mid_Slope_Positon + MRVBF + NDVI +
TWI, cDat, width = 200)

mod <- vgm(psill = var(cDat$pH60_100cm), "Sph",
range = 10000, nugget = 0)
model_1 <- fit.variogram(vgm1, mod)

gUK <- gstat(NULL, "hunterpH_UK", pH60_100cm ~ AACN +
Landsat_Band1 + Hillshading + Mid_Slope_Positon + MRVBF + NDVI +
TWI, cDat, model = model_1)
gUK
```

```
## data:
## hunterpH_UK : formula = pH60_100cm`~`AACN + Landsat_Band1 +
## Hillshading + Mid_Slope_Positon + MRVBF + NDVI + TWI ;
## data dim = 354 x 12
## variograms:
##                  model     psill    range
## hunterpH_UK[1]     Nug 0.8895319    0.000
## hunterpH_UK[2]     Sph 0.5204803 1100.581
```

7.1.2 Spatial Mapping

Here we want to produce four maps that will correspond to:

1. The lower end of the 90 % prediction interval or 5th percentile.
2. the universal kriging prediction.

3. The upper end of the 90 % prediction interval or 95th percentile.
4. The prediction interval range.

For the prediction we use the raster `interpolate` function.

```
UK.P.map <- interpolate(hunterCovariates_sub, gUK, xyOnly = FALSE,
index = 1, filename = "UK_predMap.tif",
format = "GTiff", overwrite = T)
```

Setting the `index` value to 2 lets us map the kriging variance which is needed for the prediction interval. Taking the square root this estimates the standard deviation which we can then multiple for the z value that corresponds to a 90 % probability which is `1.644854`. We then both add and subtract that result from the universal kriging prediction to derive the 90 % prediction limits.

```
# prediction variance
UK.var.map <- interpolate(hunterCovariates_sub, gUK,
xyOnly = FALSE, index = 2, filename = "UK_predVarMap.tif",
format = "GTiff", overwrite = T)

# standard deviation
f2 <- function(x) (sqrt(x))
UK.stdev.map <- calc(UK.var.map, fun = f2,
filename = "UK_predSDMap.tif", format = "GTiff",
progress = "text", overwrite = T)

# Z level
zlev <- qnorm(0.95)
f2 <- function(x) (x * zlev)
UK.mult.map <- calc(UK.stdev.map, fun = f2,
filename = "UK_multMap.tif", format = "GTiff",
progress = "text", overwrite = T)

# Add and subtract mult from prediction
m1 <- stack(UK.P.map, UK.mult.map)

# upper PL
f3 <- function(x) (x[1] + x[2])
UK.upper.map <- calc(m1, fun = f3,
filename = "UK_upperMap.tif", format = "GTiff",
progress = "text", overwrite = T)

# lower PL
f4 <- function(x) (x[1] - x[2])
UK.lower.map <- calc(m1, fun = f4, filename = "UK_lowerMap.tif",
format = "GTiff", progress = "text", overwrite = T)
```

Finally to derive the 90 % prediction limit range

```
# prediction range
m2 <- stack(UK.upper.map, UK.lower.map)

f5 <- function(x) (x[1] - x[2])
UK.piRange.map <- calc(m2, fun = f5,
filename = "UK_piRangeMap.tif", format = "GTiff",
progress = "text", overwrite = T)
```

So to plot them all together we use the following script. Here we explicitly create a color ramp that follows reasonably closely the pH color ramp. Then we scale each map to the common range for better comparison (Fig. 7.1).

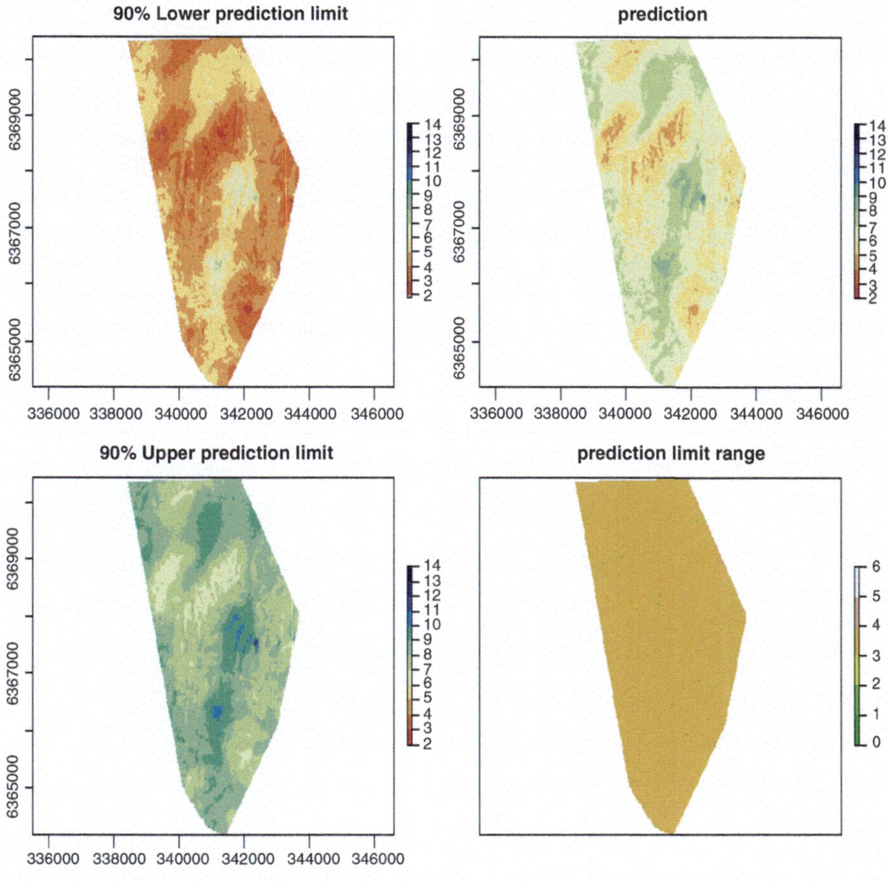

Fig. 7.1 Soil pH predictions and prediction limits derived using a universal kriging model

```
# color ramp
phCramp <- c("#d53e4f", "#f46d43", "#fdae61", "#fee08b",
    "#ffffbf", "#e6f598", "#abdda4", "#66c2a5", "#3288bd",
    "#5e4fa2", "#542788", "#2d004b")
brk <- c(2:14)
par(mfrow = c(2, 2))
plot(UK.lower.map, main = "90% Lower prediction limit",
breaks = brk, col = phCramp)
plot(UK.P.map, main = "Prediction", breaks = brk, col = phCramp)
plot(UK.upper.map, main = "90% Upper prediction limit",
breaks = brk, col = phCramp)
plot(UK.piRange.map, main = "Prediction limit range",
col = terrain.colors(length(seq(0,
6.5, by = 1)) - 1), axes = FALSE, breaks = seq(0, 6.5, by = 1))
```

7.1.3 Validating the Quantification of Uncertainty

One of the ways to assess the performance of the uncertainty quantification is to evaluate the occurrence of times where an observed value is encapsulated by an associated prediction interval. Given a stated level of confidence, we should also expect to find the same percentage of observations encapsulated by its associated prediction interval. We define this percentage as the prediction interval coverage probability (PICP). The PICP was used in both Solomatine and Shrestha (2009) and Malone et al. (2011). To assess the sensitivity of the uncertainty quantification, we define prediction intervals at a number of levels of confidence and then assess the PICP. Ideally, a 1:1 relationship would ensue.

First we apply the universal kriging model gUK to the validation data in order to estimate pH and the prediction variance.

```
coordinates(vDat) <- ~X + Y

# Prediction
UK.preds.V <- as.data.frame(krige(pH60_100cm ~ AACN +
Landsat_Band1 + Hillshading + Mid_Slope_Positon + MRVBF + NDVI +
TWI, cDat, model = model_1, newdata = vDat))

## [using universal kriging]

UK.preds.V$stdev <- sqrt(UK.preds.V$var1.var)
str(UK.preds.V)

## 'data.frame': 152 obs. of  5 variables:
##  $ X        : num  340559 340780 340861 340905 341131 ...
##  $ Y        : num  6369168 6369166 6368874 6368790 6368945 ...
##  $ var1.pred: num  7.02 7.63 6.59 5.85 5.76 ...
##  $ var1.var : num  1.13 1.18 1.15 1.16 1.12 ...
##  $ stdev    : num  1.06 1.08 1.07 1.07 1.06 ...
```

Then we define a vector of z values for a sequence of probabilities using the `qnorm` function.

```
qp <- qnorm(c(0.995, 0.9875, 0.975, 0.95, 0.9, 0.8, 0.7,
    0.6, 0.55, 0.525))
```

Then we estimate the prediction limits for each confidence level.

```
# zfactor multiplication
vMat <- matrix(NA, nrow = nrow(UK.preds.V), ncol = length(qp))
for (i in 1:length(qp)) {
    vMat[, i] <- UK.preds.V$stdev * qp[i]
}

# upper
uMat <- matrix(NA, nrow = nrow(UK.preds.V), ncol = length(qp))
for (i in 1:length(qp)) {
    uMat[, i] <- UK.preds.V$var1.pred + vMat[, i]
}

# lower
lMat <- matrix(NA, nrow = nrow(UK.preds.V), ncol = length(qp))
for (i in 1:length(qp)) {
    lMat[, i] <- UK.preds.V$var1.pred - vMat[, i]
}
```

Then we want to evaluate the PICP for each confidence level.

```
bMat <- matrix(NA, nrow = nrow(UK.preds.V), ncol = length(qp))
for (i in 1:ncol(bMat)) {
    bMat[, i] <- as.numeric(vDat$pH60_100cm <= uMat[, i] &
    vDat$pH60_100cm >= lMat[, i])
}

colSums(bMat)/nrow(bMat)

##  [1] 0.98026316 0.96052632 0.94078947 0.88157895
##  0.80921053 0.63815789
##  [7] 0.46052632 0.25000000 0.07236842 0.03947368
```

Plotting the confidence level against the PICP provides a visual means to assess the fidelity about the 1:1 line. As can be seen on Fig. 7.2, the PICP follows closely the 1:1 line.

```
# make plot
cs <- c(99, 97.5, 95, 90, 80, 60, 40, 20, 10, 5)
plot(cs, ((colSums(bMat)/nrow(bMat)) * 100))
```

So to summarize. We may evaluate the performance of the universal kriging model on the basis of the predictions. Using the validation data we would use the `goof` function for that purpose.

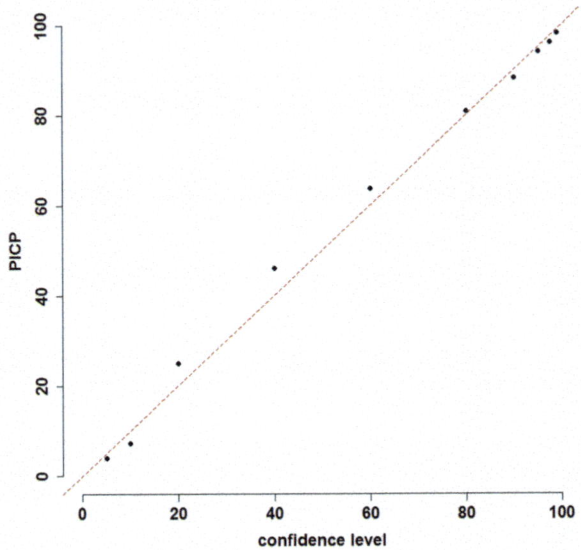

Fig. 7.2 Plot of PICP and confidence level based on validation of universal kriging model

```
goof(observed = vDat$pH60_100cm, predicted = UK.preds.V$var1.pred)

##          R2 concordance      MSE     RMSE      bias
## 1 0.3628927   0.5303003 1.158743 1.076449 0.1709653
```

And then we may assess the uncertainties on the basis of the PICP like shown on Fig. 7.2, together with assessing the quantiles of the distribution of the prediction limit range for a given prediction confidence level (here 90 %).

```
cs <- c(99, 97.5, 95, 90, 80, 60, 40, 20, 10, 5)
colnames(lMat) <- cs
colnames(uMat) <- cs
quantile(uMat[, "90"] - lMat[, "90"])

##       0%      25%      50%      75%     100%
## 3.293106 3.410418 3.461094 3.532295 3.866376
```

As can be noted above, the prediction interval range is relatively homogeneous and this is corroborated on the associated map in Fig. 7.1.

7.2 Bootstrapping

Bootstrapping is a popular non-parametric approach for quantifying prediction uncertainties (Efron and Tibshirani 1993). Bootstrapping involves repeated random sampling with replacement of the available data. With the bootstrap sample, a model is fitted, and can then be applied to generate a digital soil map. By repeating the process of random sampling and applying the model, we are able to generate probability distributions of the prediction realizations from each model at each pixel. A robust estimate may be determined by taking the average of all the simulated predictions at each pixel. By being able to obtain probability distributions of the outcomes, one is also able to quantify the uncertainty of the modeling by computing a prediction interval given a specified level of confidence. While the bootstrapping approach is relatively straightforward, there is a requirement to generate x number of maps, where x is the number of bootstrap samples. This obviously could be prohibitive from a computational and data storage point of view, but not altogether impossible (given parallel processing capabilities etc.) as was demonstrated by both Viscarra Rossel et al. (2015) and Liddicoat et al. (2015) whom both performed bootstrapping for quantification of uncertainties across very large mapping extents. In the case of Viscarra Rossel et al. (2015) this was for the entire Australian continent at 100 m resolution.

In the example below, the bootstrap method is demonstrated. We will be using Cubist modeling for the model structure and perform 50 bootstrap samples. We will use 70 % of the available data to use for fitting models. The remaining 30 % as has been done for all previous DSM approaches is for validation.

7.2.1 Defining the Model Parameters

For the first step, we do the random partitioning of the data into calibration and validation data sets. Again we are using the `HV_subsoilpH` data and the associated `hunterCovariates_sub` raster data stack.

```
# subset data for modeling
set.seed(667)
training <- sample(nrow(HV_subsoilpH), 0.7 * nrow(HV_subsoilpH))
cDat <- HV_subsoilpH[training, ]
vDat <- HV_subsoilpH[-training, ]
```

The `nbag` variable below holds the value for the number of bootstrap models we want to fit. Here it is 50. Essentially the bootstrap can can be contained within a `for` loop, where upon each loop a sample of the available data is taken (here 100 %) then a model is fitted. Note below the use of the `replace` parameter to indicate we want random sample with replacement. After a model is fitted, we save the model to file and will come back to it later. The `modelFile` variable shows the extensive use of the `paste` function in order to provide the pathway and file name for the model that we want to save on each loop iteration. The `saveRDS` function allows us to save each of the model objects as rds files to the location specified. An alternative to save the models individually to file is to save them to elements within a `list`. When dealing with very large numbers of models and additionally are complex in terms of their parameterizations, the save to `list` elements alternative could run into computer memory limitation issues. The last section of the script below just demonstrates the use of the `list.files` functions to confirm that we have saved those models to file and they are ready to use.

```
# Number of bootstraps
nbag <- 50

# Fit cubist models for each bootstrap
library(Cubist)
for (i in 1:nbag) {
    trainingREP <- sample.int(nrow(cDat), 1.0 * nrow(cDat),
        replace = TRUE)

    fit_cubist <- cubist(x = cDat[trainingREP,
        c("Terrain_Ruggedness_Index",
         "AACN", "Landsat_Band1", "Elevation", "Hillshading",
         "Light_insolation", "Mid_Slope_Positon", "MRVBF", "NDVI",
         "TWI", "Slope")],
         y = cDat$pH60_100cm[trainingREP], cubistControl(rules = 5,
         extrapolation = 5), committees = 1)

    modelFile <- paste(paste(paste(paste(getwd(),
        "~~/bootstrap/models/", sep = ""), "bootMod_",
             sep = ""), i, sep = ""), ".rds", sep = "")

    saveRDS(object = fit_cubist, file = modelFile)
}

# list all files in directory
c.models <- list.files(path = paste(getwd(),"~~/bootstrap/models",
    sep = ""), pattern = "\\.rds$", full.names = TRUE)
head(c.models)

## [1] "~~/bootstrap/models/bootMod_1.rds"
## [2] "~~/bootstrap/models/bootMod_10.rds"
## [3] "~~/bootstrap/models/bootMod_11.rds"
## [4] "~~/bootstrap/models/bootMod_12.rds"
## [5] "~~/bootstrap/models/bootMod_13.rds"
## [6] "~~/bootstrap/models/bootMod_14.rds"
```

We can then assess the goodness of fit and validation statistics of the bootstrap models. This is done using the `goof` function as in previous examples. This time we incorporate that function within a `for` loop. For each loop, we read in the model via the `radRDS` function and then save the diagnostics to the `cubiMat` matrix object. After the iterations are completed, we use the `colMeans` function to calculate the means of the diagnostics over the 50 model iterations. You could also assess the variance of those means by a command such as `var(cubiDat[,1])`, which would return the variance of the R^2 values.

```
# calibration data
cubiMat <- matrix(NA, nrow = nbag, ncol = 5)
for (i in 1:nbag) {
    fit_cubist <- readRDS(c.models[i])
    cubiMat[i, ] <- as.matrix(goof(observed = cDat$pH60_100cm,
    predicted = predict(fit_cubist, newdata = cDat)))
}

cubiDat <- as.data.frame(cubiMat)
names(cubiDat) <- c("R2", "concordance", "MSE", "RMSE", "bias")
colMeans(cubiDat)

##          R2 concordance         MSE        RMSE        bias
##  0.25261147  0.45185565  1.46214146  1.20887737 -0.06697598

# Validation data
cubPred.V <- matrix(NA, ncol = nbag, nrow = nrow(vDat))
cubiMat <- matrix(NA, nrow = nbag, ncol = 5)
for (i in 1:nbag) {
    fit_cubist <- readRDS(c.models[i])
    cubPred.V[, i] <- predict(fit_cubist, newdata = vDat)
    cubiMat[i, ] <- as.matrix(goof(observed = vDat$pH60_100cm,
    predicted = predict(fit_cubist, newdata = vDat)))
}
cubPred.V_mean <- rowMeans(cubPred.V)

cubiDat <- as.data.frame(cubiMat)
names(cubiDat) <- c("R2", "concordance", "MSE", "RMSE", "bias")
colMeans(cubiDat)

##          R2 concordance         MSE        RMSE        bias
##  0.09010054  0.26690013  1.80777927  1.34013203  0.11262625

# Average validation MSE
avGMSE <- mean(cubiDat[, 3])
```

For the validation data, in addition to deriving the model diagnostic statistics, we are also saving the actual model predictions for these data for each iteration to the `cubPred.V` object. These will be used further on for validating the prediction uncertainties. The last line of the script above saves the mean of the mean square error (MSE) estimates from the validation data. The independent MSE estimator, accounts for both systematic and random errors in the modeling. This estimate of the MSE is needed for quantifying the uncertainties, as this error is in addition to that which are accounted for by the bootstrap, which are specifically those associated with the deterministic model component i.e. the model relationship between target variable and the covariates. Subsequently an overall prediction variance (at each point or pixel) will be the sum of the random error component (MSE) and the bootstrap prediction variance (as estimated from the mean of the realisations from the bootstrap modeling).

7.2.2 Spatial Mapping

Our initial purpose here is to derive the mean and the variance of the predictions from each bootstrap sample. This requires loading in each bootstrap model, applying into the covariate data, then saving the predicted map to file or R memory. In the case below the predictions are saved to file. This is illustrated in the following script.

```
for (i in 1:nbag) {
    fit_cubist <- readRDS(c.models[i])
    mapFile <- paste(paste(paste(paste(getwd(), "~~/bootstrap/map/",
    sep = ""), "bootMap_", sep = ""), i, sep = ""), ".tif", sep = "")
    predict(hunterCovariates_sub, fit_cubist, filename = mapFile,
    format = "GTiff", overwrite = T)
}
```

To evaluate the mean at each pixel from each of the created maps, the base function `mean` can be applied to a given stack of rasters. First we need to get the path location of the rasters. Notice from the `list.files` function and the `pattern` parameter, we are restricting the search of rasters that contain the string "bootMap". Next we make a stack of those rasters, followed by the calculation of the mean, which is also written directly to file.

```
# Pathway to rasters
files <- list.files(paste(getwd(), "~~/bootstrap/map/", sep = ""),
pattern = "bootMap", full.names = TRUE)

# Raster stack
r1 <- raster(files[1])
for (i in 2:length(files)) {
    r1 <- stack(r1, files[i])
}

# Calculate mean
meanFile <- paste(paste(paste(getwd(), "~~/bootstrap/map/",
sep = ""), "meanPred_", sep = ""), ".tif", sep = "")
bootMap.mean <- writeRaster(mean(r1), filename = meanFile,
  format = "GTiff", overwrite = TRUE)
```

There is not a simple `R` function to use in order to estimate the variance at each pixel from the prediction maps. Therefore we resort to estimating it directly from the standard equation:

$$Var(X) = \frac{1}{1-n}\sum_{i=1}^{n}(x_i - \mu)^2 \tag{7.1}$$

The symbol μ in this case is the mean bootstrap prediction, and x_i is the ith bootstrap map. In the first step below, we estimate the square differences and save the maps to file. Then we calculate the sum of those squared differences, before deriving the variance prediction. The last step is to add the variance of the bootstrap predictions to the averaged MSE estimated from the validation data.

```
# Square differences
for (i in 1:length(files)) {
    r1 <- raster(files[i])
    diffFile <- paste(paste(paste(paste(getwd(),
       "~~/bootstrap/map/",
    sep = ""), "bootAbsDif_", sep = ""), i, sep = ""),
        ".tif", sep = "")
    jj <- (r1 - bootMap.mean)^2
    writeRaster(jj, filename = diffFile, format = "GTiff",
    overwrite = TRUE)
}

# calculate the sum of square differences (Look for files with
  the bootAbsDif)
# character string in file name
files2 <- list.files(paste(getwd(), "~~/bootstrap/map/", sep =""),
pattern = "bootAbsDif", full.names = TRUE)

# stack
r2 <- raster(files2[1])
for (i in 2:length(files2)) {
    r2 <- stack(r1, files2[i])
}

sqDiffFile <- paste(paste(paste(getwd(), "~~/bootstrap/map/",
   sep = ""),
"sqDiffPred_", sep = ""), ".tif", sep = "")
bootMap.sqDiff <- writeRaster(sum(r2), filename = sqDiffFile,
format = "GTiff", overwrite = TRUE)

# Variance
varFile <- paste(paste(paste(getwd(), "~~/bootstrap/map/", sep=""),
"varPred_", sep = ""), ".tif", sep = "")
bootMap.var <- writeRaster(((1/(nbag - 1)) * bootMap.sqDiff),
filename = varFile, format = "GTiff", overwrite = TRUE)

# Overall prediction variance
varFile2 <- paste(paste(paste(getwd(), "~~/bootstrap/map/",
   sep=""),
"varPredF_", sep=""), ".tif",
sep="")
bootMap.varF <- writeRaster((bootMap.var + avGMSE),
filename = varFile,
format = "GTiff", overwrite = TRUE)
```

To derive to 90 % prediction interval we take the square root of the variance estimate and multiply that value by the z value that corresponds to a 90 %

probability. The z value is obtained using the qnorm function. The result is then either added or subtracted to the mean prediction in order to generate the upper and lower prediction limits respectively.

```
# Standard deviation
sdFile <- paste(paste(paste(getwd(), "~~/bootstrap/map/",
sep = ""), "sdPred_", sep = ""), ".tif", sep = "")
bootMap.sd <- writeRaster(sqrt(bootMap.varF), filename = sdFile,
format = "GTiff", overwrite = TRUE)

# standard error
seFile <- paste(paste(paste(getwd(), "~~/bootstrap/map/",
sep = ""), "sePred_", sep = ""), ".tif", sep = "")
bootMap.se <- writeRaster((bootMap.sd * qnorm(0.95)),
filename = seFile, format = "GTiff", overwrite = TRUE)

# upper prediction limit
uplFile <- paste(paste(paste(getwd(), "~~/bootstrap/map/",
sep = ""), "uplPred_", sep = ""), ".tif", sep = "")
bootMap.upl <- writeRaster((bootMap.mean + bootMap.se),
filename = uplFile, format = "GTiff", overwrite = TRUE)

# lower prediction limit
lplFile <- paste(paste(paste(getwd(), "~~/bootstrap/map/",
sep = ""), "lplPred_", sep = ""), ".tif", sep = "")
bootMap.lpl <- writeRaster((bootMap.mean - bootMap.se),
filename = lplFile, format = "GTiff", overwrite = TRUE)

# prediction interval range
pirFile <- paste(paste(paste(getwd(), "~~/bootstrap/map/",
sep = ""), "pirPred_", sep = ""), ".tif", sep = "")
bootMap.pir <- writeRaster((bootMap.upl - bootMap.lpl),
filename = pirFile, format = "GTiff", overwrite = TRUE)
```

As for the Universal kriging example, we can plot the associated maps of the predictions and quantified uncertainties (Fig. 7.3).

```
phCramp <- c("#d53e4f", "#f46d43", "#fdae61", "#fee08b",
"#ffffbf", "#e6f598", "#abdda4", "#66c2a5", "#3288bd", "#5e4fa2",
"#542788", "#2d004b")
brk <- c(2:14)
par(mfrow = c(2, 2))
plot(bootMap.lpl, main = "90% Lower prediction limit",
breaks = brk, col = phCramp)
plot(bootMap.mean, main = "Prediction",
breaks = brk, col = phCramp)
plot(bootMap.upl, main = "90% Upper prediction limit",
breaks = brk, col = phCramp)
plot(bootMap.pir, main = "Prediction limit range",
col = terrain.colors(length(seq(0,
6.5, by = 1)) - 1), axes = FALSE, breaks = seq(0, 6.5, by = 1))
```

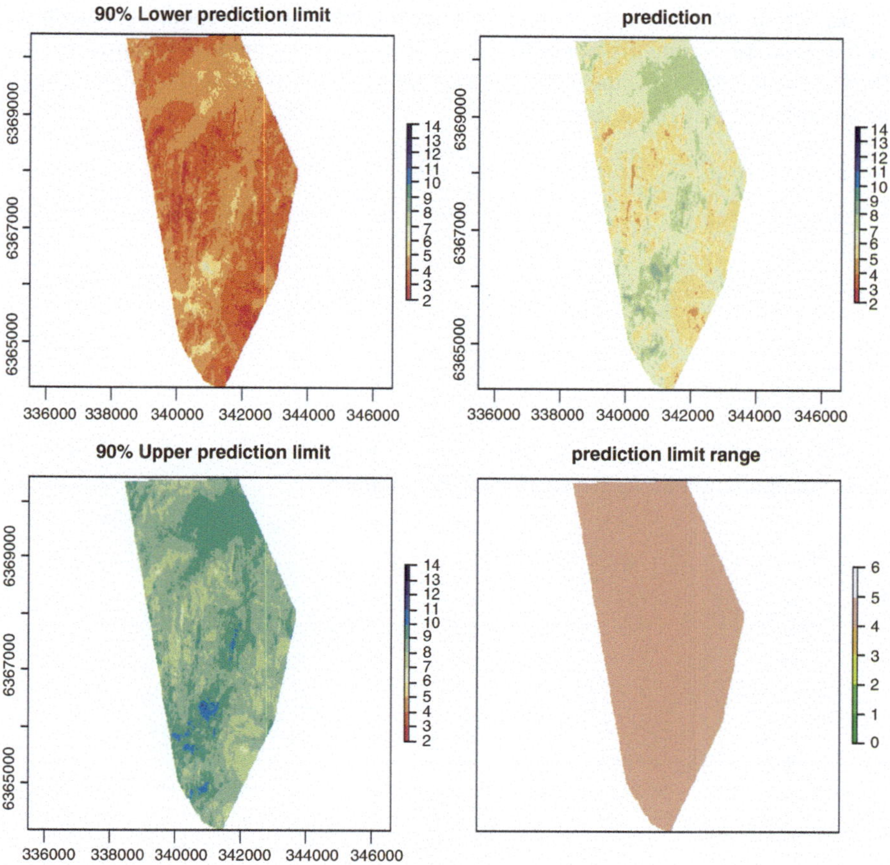

Fig. 7.3 Soil pH predictions and prediction limits derived using bootstrapping

7.2.3 *Validating the Quantification of Uncertainty*

You will recall the bootstrap model predictions on the validation data were saved to the `cubPred.V` object. We want estimate the standard deviation of those predictions for each point. Also recall that the prediction variance is the sum of the MSE and the bootstrap models prediction variance. Taking the square root of that summation results in standard deviation estimate.

```
val.sd <- matrix(NA, ncol = 1, nrow = nrow(cubPred.V))
for (i in 1:nrow(cubPred.V)) {
    val.sd[i, 1] <- sqrt(var(cubPred.V[i, ]) + avGMSE)
}
```

We then need to multiply the standard deviation by the corresponding percentile of the standard normal distribution in order to express the prediction limits at each level of confidence. Note the use of the `for` loop and the associated cycling through of the different percentile values.

```
# Percentiles of normal distribution
qp <- qnorm(c(0.995, 0.9875, 0.975, 0.95, 0.9, 0.8, 0.7, 0.6,
   0.55, 0.525))

# zfactor multiplication
vMat <- matrix(NA, nrow = nrow(cubPred.V), ncol = length(qp))
for (i in 1:length(qp)) {
    vMat[, i] <- val.sd * qp[i]
}
```

Now we add or subtract the limits to/from the averaged model predictions to derive to prediction limits for each level of confidence.

```
# upper prediction limit
uMat <- matrix(NA, nrow = nrow(cubPred.V), ncol = length(qp))
for (i in 1:length(qp)) {
    uMat[, i] <- cubPred.V_mean + vMat[, i]
}

# lower prediction limit
lMat <- matrix(NA, nrow = nrow(cubPred.V), ncol = length(qp))
for (i in 1:length(qp)) {
    lMat[, i] <- cubPred.V_mean - vMat[, i]
}
```

Now we assess the PICP for each level confidence. Recalling that we are simply assessing whether the observed value is encapsulated by the corresponding prediction limits, then calculating the proportion of agreement to total number of observations.

```
bMat <- matrix(NA, nrow = nrow(cubPred.V), ncol = length(qp))
for (i in 1:ncol(bMat)) {
    bMat[, i] <- as.numeric(vDat$pH60_100cm <= uMat[,i]
    & vDat$pH60_100cm >= lMat[, i])
}

colSums(bMat)/nrow(bMat)

##  [1] 1.00000000 1.00000000 0.99342105 0.96710526 0.90131579
    0.69078947
##  [7] 0.45394737 0.21052632 0.08552632 0.03289474
```

As can be seen on Fig. 7.4, there is an indication that the prediction uncertainties could be a little too liberally defined, where particularly at the higher level of confidence the associated PICP is higher.

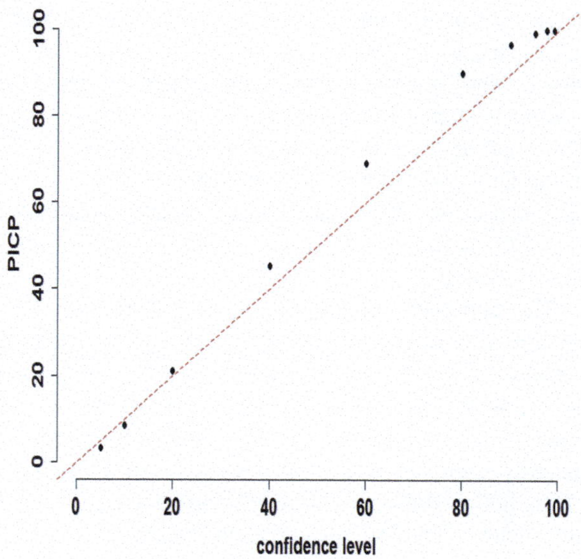

Fig. 7.4 Plot of PICP and confidence level based on validation of bootstrapping model

```
# make plot
cs <- c(99, 97.5, 95, 90, 80, 60, 40, 20, 10, 5)
plot(cs, ((colSums(bMat)/nrow(bMat)) * 100))
```

Quantiles of the distribution of the prediction limit range are express below for the validation data (in terms of the 90 % level of confidence). Compared to the universal kriging approach, the uncertainties quantified from the bootstrapping approach are higher in general.

```
cs <- c(99, 97.5, 95, 90, 80, 60, 40, 20, 10, 5)
colnames(lMat) <- cs
colnames(uMat) <- cs
quantile(uMat[, "90"] - lMat[, "90"])

##       0%      25%      50%      75%     100%
## 4.551972 4.702860 4.795511 4.950161 6.610331
```

7.3 Empirical Uncertainty Quantification Through Data Partitioning and Cross Validation

The next two approaches for uncertainty quantification are empirical methods whereby the prediction intervals are defined from the distribution of model errors i.e. the deviation between observation and model predictions. In the examples to follow however, the prediction limits are not spatially uniform. Rather they are a function of the landscape. For example, in some areas of the landscape, or in some particular landscape situations, the prediction is going to be more accurate than in other situations. For the immediate approach, an explicit landscape partition is used to define likely areas where the quantification of uncertainty may differ. This approach uses the Cubist data mining algorithm for prediction and partitioning of the landscape. Within each partitioned area, we can then define the distribution of model errors. This approach for uncertainty quantification was used by Malone et al. (2014) where soil pH was predicted using a Cubist model and residual kriging spatial model.

This approach and the next are useful from the viewpoint that complex models (in terms of parameterisation) are able to be used within a digital soil mapping framework where there is a necessity to define a stated level of prediction confidence. With universal kriging, this is the restriction of the linear model (which in many situations could be the best model). For empirical approaches to uncertainty quantification, this restriction does not exist. Further to this, the localized allocation of prediction limits i.e. in given landscape situations provides a means to identify areas where mapping is accurate and where it is not so much. Often this is a function of the sampling density. This explicit allocation of prediction intervals may assist with the future design of soil sampling programs.

In the approach to be described the partitioning is a “hard” partition, for the next approach described in the next section, it is a “soft” or “fuzzy” partition.

7.3.1 Defining the Model Parameters

We will use a Cubist model regression kriging approach as used previously, and then define the subsequent prediction intervals.

First we prepare the data.

```
# subset data for modeling
set.seed(667)
training <- sample(nrow(HV_subsoilpH), 0.7 * nrow(HV_subsoilpH))
cDat <- HV_subsoilpH[training, ]
vDat <- HV_subsoilpH[-training, ]
```

Now we fit the Cubist model using all available environmental covariates.

```
library(Cubist)
# Cubist model
hv.cub.Exp <- cubist(x = cDat[, c("Terrain_Ruggedness_Index",
"AACN", "Landsat_Band1", "Elevation", "Hillshading",
"Light_insolation", "Mid_Slope_Positon", "MRVBF", "NDVI", "TWI",
"Slope")],
y = cDat$pH60_100cm, cubistControl(unbiased =TRUE, rules = 100,
extrapolation = 10, sample = 0, label =  "outcome"), committees=1)
summary(hv.cub.Exp)
```

```
##
## Read 354 cases (12 attributes) from undefined.data
##
## Model:
##
##   Rule 1: [116 cases, mean 6.183380, range 3.956412 to
            9.249626, est err 0.714710]
##
##     if
##  NDVI <= -0.191489
##  TWI <= 13.17387
##     then
##  outcome = 9.610208 + 0.0548 AACN - 0.0335 Elevation + 0.131
##            Hillshading + 3.5 NDVI + 0.076 Terrain_Ruggedness_Index
##            - 0.055 Slope - 0.023 Landsat_Band1 + 0.03 MRVBF
##            + 0.07 Mid_Slope_Positon + 0.005 TWI
##
##   Rule 2: [164 cases, mean 6.533212, range 3.437355 to 9.741758,
            est err 0.986175]
##
##     if
##  TWI > 13.17387
##     then
##  outcome = 7.471082 + 0.0215 AACN + 0.108 Hillshading + 4.2 NDVI
##            + 0.24 MRVBF + 1.16 Mid_Slope_Positon - 0.0104 Elevation
##            - 0.069 Slope + 0.077 Terrain_Ruggedness_Index
##            - 0.028 Landsat_Band1 + 0.047 TWI
##
##   Rule 3: [74 cases, mean 6.926269, range 2.997182 to 9.630296,
            est err 1.115631]
##
##     if
##  NDVI > -0.191489
##  TWI <= 13.17387
##     then
##  outcome = 11.743466 + 0.0416 AACN - 0.091 Landsat_Band1
##            - 0.0117 Elevation + 3 NDVI + 0.048 Hillshading
##            + 0.065 Terrain_Ruggedness_Index - 0.046 Slope
##
##
## Evaluation on training data (354 cases):
##
##     Average  |error|              1.085791
##     Relative |error|                  0.95
##     Correlation coefficient           0.4
```

The `hv.cub.Exp` model indicates three subsets in the data were made. Assessing the goodness of fit of this model, we use the `goof` function. This model seems to perform OK.

```
goof(observed = cDat$pH60_100cm, predicted = predict(hv.cub.Exp,
  newdata = cDat))
```

```
##          R2 concordance       MSE     RMSE          bias
## 1 0.3247524   0.4779242 1.266316 1.125307 2.640776e-07
```

Now we want to assess the model residuals for spatial auto correlation. In this case we need to look at the variogram of the residuals.

```
cDat$residual <- cDat$pH60_100cm - predict(hv.cub.Exp,
   newdata = cDat)

# coordinates
coordinates(cDat) <- ~X + Y
# residual variogram model
vgm1 <- variogram(residual ~ 1, cDat, width = 200)
mod <- vgm(psill = var(cDat$residual), "Sph", range = 10000,
   nugget = 0)
model_1 <- fit.variogram(vgm1, mod)

gOK <- gstat(NULL, "hunterpH_cubistRES", residual ~ 1, cDat,
   model = model_1)
gOK

## data:
## hunterpH_cubistRES : formula = residual`~`1 ; data dim = 354 x 13
## variograms:
##                        model     psill    range
## hunterpH_cubistRES[1]    Nug 0.7431729   0.0000
## hunterpH_cubistRES[2]    Sph 0.5385914 937.8323
```

This output indicates there is a reasonable variogram of the residuals with which may help improve the overall predictive model. We can determine whether there is any improvement later when we perform the validation.

With a model defined, it is now necessary to estimate the within partition prediction limits. This is done using a leave-one-out cross validation procedure using a Cubist model regression kriging model as we did for the spatial model. What we need to do first is to determine which observations in the data set belong to which partition. Looking at the `summary` output above we can examine the threshold criteria that define the data partitions.

```
# Assign a rule to each observation if more than one observation
cDat1 <- as.data.frame(cDat)
cDat1$rule = 9999
# rule 1
cDat1$rule[which(cDat1$NDVI <= -0.191489 & cDat1$TWI<=13.17387)] <- 1
# rule 2
cDat1$rule[which(cDat1$TWI> 13.17387)] <-2
## rule 3
cDat1$rule[which(cDat1$NDVI > -0.191489 & cDat1$TWI <= 13.17387)] <-3

cDat1$rule <- as.factor(cDat1$rule)
summary(cDat1$rule)

##   1   2   3
## 115 165  74
```

Now we can subset `cDat1` based on its defined rule or partition and then perform the LOCV. The script below shows the procedure for the process using the first rule.

```
# subset the data
cDat1.r1 <- cDat1[which(cDat1$rule == 1), ]
target.C.r1 <- cDat1.r1$pH60_100cm

########## Leave-one-out cross validation ###################
looResiduals <- numeric(nrow(cDat1.r1))
for (i in 1:nrow(cDat1.r1)) {
    # fit cubist model
    loocubistPred <- cubist(x = cDat1.r1[-i, 4:14], y=target.C.r1[-i],
    cubistControl(unbiased = F, rules = 1, extrapolation = 5,
    sample = 0, seed = sample.int(4096, size = 1) - 1L,
    label = "outcome"), committees = 1)
    cDat11.r1.sub <- cDat1.r1[-i, ]
    cDat11.r1.sub$pred <- predict(loocubistPred,
    newdata = cDat11.r1.sub,
    neighbors = 0)
    cDat11.r1.sub$resids <- target.C.r1[-i] - cDat11.r1.sub$pred
    # residual variogram
    vgm.r1 <- variogram(resids ~ 1, ~X + Y, cDat11.r1.sub,
    width = 100)
    mod.r1 <- vgm(psill = var(cDat11.r1.sub$resids), "Sph",
    range = 10000, nugget = 0)
    model_1.r1 <- fit.variogram(vgm.r1, mod.r1)
    # interpolate residual on withheld point
    int.resids1.r1 <- krige(cDat11.r1.sub$resids ~ 1,
    locations = ~X + Y, data = cDat11.r1.sub, newdata
    = cDat1.r1[i, c("X", "Y")],
    model = model_1.r1)[, 3]
    # Cubist model predict on withheld point
    looPred <- predict(loocubistPred, newdata = cDat1.r1[i, ],
    neighbors = 0)
    # Add cubist model prediction and residual
    looResiduals[i] <- target.C.r1[i] - (looPred + int.resids1.r1)
}
```

We are interested in evaluating the 90 % prediction interval. We do this for each data partition by taking the lower 5th and upper 95th quantiles of the residual distribution, and ultimately add these to the model predictions. For the validation, we also have to evaluate the quantiles for a range of confidence levels (in this case sequentially from 5 % to 99 %) for calculation of the PICP. This is done using the `quantile` function.

```
# Rule 1 90% confidence
r1.ulPI <- quantile(looResiduals1, probs = c(0.05, 0.95),
na.rm = FALSE, names = F, type = 7)
r1.ulPI

## [1] -1.023353  1.433298
```

```
# Confidence interval range
r1.q <- quantile(looResiduals1, probs = c(0.005, 0.995, 0.0125,
0.9875, 0.025, 0.975, 0.05, 0.95, 0.1, 0.9, 0.2, 0.8, 0.3, 0.7,
0.4, 0.6, 0.45, 0.55, 0.475, 0.525),
na.rm = FALSE, names = F, type = 7)

# Rule 2 90% confidence
r2.ulPI <- quantile(looResiduals2, probs = c(0.05, 0.95),
na.rm = FALSE, names = F, type = 7)
r2.ulPI

## [1] -2.416407  2.187128

# Confidence interval range
r2.q <- quantile(looResiduals2, probs = c(0.005, 0.995, 0.0125,
0.9875, 0.025, 0.975, 0.05, 0.95, 0.1, 0.9, 0.2, 0.8,0.3, 0.7,
0.4, 0.6, 0.45, 0.55, 0.475, 0.525), na.rm = FALSE,
names = F, type = 7)

# Rule 3 90% confidence
r3.ulPI <- quantile(looResiduals3, probs = c(0.05, 0.95),
na.rm = FALSE, names = F, type = 7)
r3.ulPI

## [1] -1.629060  1.732905

# Confidence interval range
r3.q <- quantile(looResiduals3, probs = c(0.005, 0.995, 0.0125,
0.9875, 0.025, 0.975, 0.05, 0.95, 0.1, 0.9, 0.2, 0.8, 0.3, 0.7,
0.4, 0.6, 0.45, 0.55, 0.475, 0.525), na.rm = FALSE,
names = F, type = 7)
```

As can be seen, the prediction limits for each partition of the data are quite different from each other.

7.3.2 *Spatial Mapping*

To create the maps in the same fashion as was done for bootstrapping and universal kriging, we do more-or-less as the same for the Cubist regression kriging. First create the regression kriging map.

```
# Cubist predicted map
map.cubist <- predict(hunterCovariates_sub, hv.cub.Exp,
args = list(neighbors = 0), filename = "rk_cubist.tif",
format = "GTiff", overwrite = T)
```

```
# kriged residuals
map.cubist.res <- interpolate(hunterCovariates_sub, gOK,
xyOnly = TRUE, index = 1, filename = "rk_residuals.tif",
format = "GTiff", datatype = "FLT4S",  overwrite = TRUE)

# raster stack of predictions and residuals
r2 <- stack(map.cubist, map.cubist.res)
f1 <- function(x) calc(x, sum)
map.cubist.final <- calc(r2, fun = sum, filename = "cubistRK.tif",
format = "GTiff", overwrite = T)
```

To derive the upper and lower prediction limits we can apply the raster calculations in a very manual way, such that each line of the `rasterStack` is read in and then evaluated to determine which rule each entry on the line belongs to. Then given the rule, the corresponding upper and lower limits are appended to a new raster together with a raster which indicates which rule was applied where. For small raster data sets where the mapping extent is small, this approach works fine. It can also be applied for very large rasters, but can take some time as the reading of the raster occurs line by line. There are however, raster options to process them this way in chunk form.

```
# Create new raster datasets
upper1 <- raster(hunterCovariates_sub[[1]])
lower1 <- raster(hunterCovariates_sub[[1]])
rule1 <- raster(hunterCovariates_sub[[1]])
upper1 <- writeStart(upper1, filename = "cubRK_upper1.tif",
format = "GTiff", overwrite = TRUE)
lower1 <- writeStart(lower1, filename = "cubRK_lower1.tif",
format = "GTiff", overwrite = TRUE)
rule1 <- writeStart(rule1, filename = "cubRK_rule1.tif",
format = "GTiff", datatype = "INT2S", overwrite = TRUE)

for (i in 1:dim(upper1)[1]) {
    # extract raster information line by line
    cov.Frame <- as.data.frame(getValues(hunterCovariates_sub, i))
    ulr.Frame <- matrix(NA, ncol = 3, nrow = dim(upper1)[2])
    # append in partition information

    # rule 1
    ulr.Frame[which(cov.Frame$NDVI <= -0.191489 & cov.Frame$TWI
         <= 13.17387), 1] <- r1.ulPI[2]
    ulr.Frame[which(cov.Frame$NDVI <= -0.191489 & cov.Frame$TWI
       <= 13.17387), 2] <- r1.ulPI[1]
    ulr.Frame[which(cov.Frame$NDVI <= -0.191489 & cov.Frame$TWI
       <= 13.17387), 3] <- 1
    # rule 2
    ulr.Frame[which(cov.Frame$TWI > 13.17387), 1] <- r2.ulPI[2]
    ulr.Frame[which(cov.Frame$TWI > 13.17387), 2] <- r2.ulPI[1]
    ulr.Frame[which(cov.Frame$TWI > 13.17387), 3] <- 2
    # rule 3
    ulr.Frame[which(cov.Frame$NDVI > -0.191489 & cov.Frame$TWI
```

```
        <= 13.17387), 1] <- r3.ulPI[2]
    ulr.Frame[which(cov.Frame$NDVI > -0.191489 & cov.Frame$TWI
        <= 13.17387), 2] <- r3.ulPI[1]
    ulr.Frame[which(cov.Frame$NDVI > -0.191489 & cov.Frame$TWI
        <= 13.17387), 3] <- 3

    ulr.Frame <- as.data.frame(ulr.Frame)
    names(ulr.Frame) <- c("upper", "lower", "rule")
    # write to raster then close
    pred_upper <- ulr.Frame$upper
    pred_lower <- ulr.Frame$lower
    pred_rule <- ulr.Frame$rule
    upper1 <- writeValues(upper1, pred_upper, i)
    lower1 <- writeValues(lower1, pred_lower, i)
    rule1 <- writeValues(rule1, pred_rule, i)
    print(i)
}
upper1 <- writeStop(upper1)
lower1 <- writeStop(lower1)
rule1 <- writeStop(rule1)
```

Now we can derive the prediction interval by adding the upper and lower limits to the regression kriging prediction that was made earlier. Then we can estimate the prediction interval range.

```
# raster stack of predictions and prediction limits
r2 <- stack(map.cubist.final, lower1)  #lower
mapRK.lower <- calc(r2, fun = sum, filename = "cubistRK_lowerPL.tif",
format = "GTiff", overwrite = T)

# raster stack of predictions and prediction limits
r2 <- stack(map.cubist.final, upper1)  #upper
mapRK.upper <- calc(r2, fun = sum, filename = "cubistRK_upperPI.tif",
format = "GTiff", overwrite = T)

# Prediction interval range
r2 <- stack(mapRK.lower, mapRK.upper)  #diff
mapRK.PIrange <- calc(r2, fun=diff, filename = "cubistRK_PIrange.tif",
format = "GTiff", overwrite = T)
```

Now we can plot the maps as before. We can note this time that the prediction interval range is smaller for the cubist regression kriging than it is for the universal kriging approach (Fig. 7.5).

```
# color ramp
phCramp <- c("#d53e4f", "#f46d43", "#fdae61", "#fee08b", "#ffffbf",
"#e6f598", "#abdda4", "#66c2a5", "#3288bd", "#5e4fa2", "#542788",
"#2d004b")
brk <- c(2:14)
par(mfrow = c(2, 2))
plot(mapRK.lower, main = "90% Lower prediction limit",
breaks = brk, col = phCramp)
```

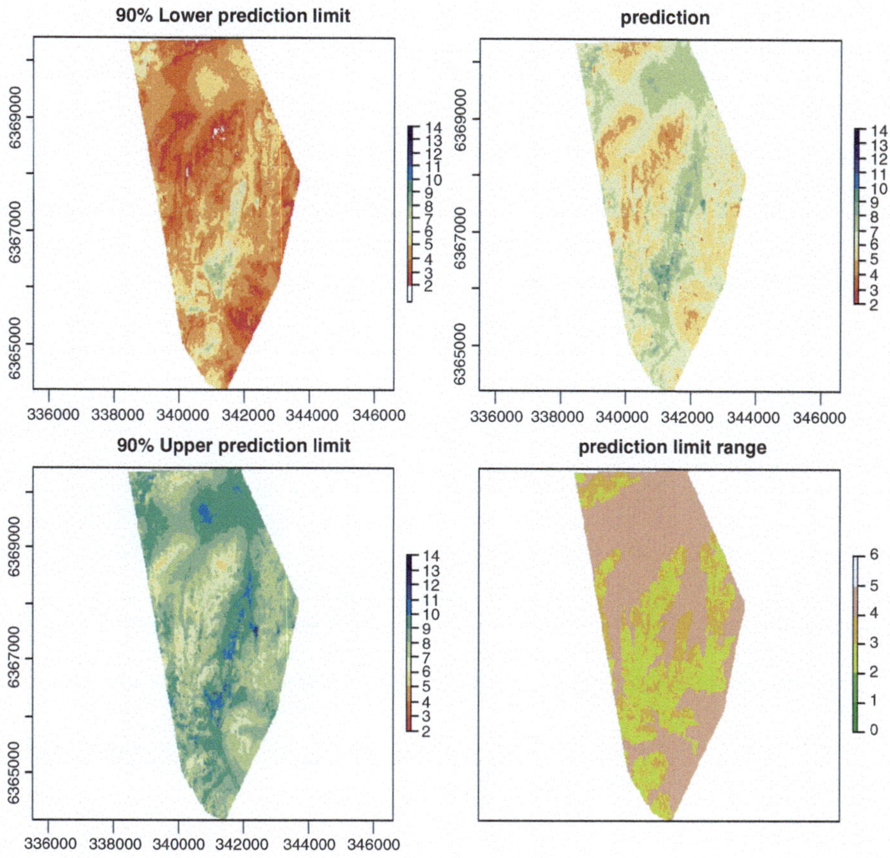

Fig. 7.5 Soil pH predictions and prediction limits derived using a Cubist regression kriging model

```
plot(map.cubist.final, main = "Prediction",
breaks = brk, col = phCramp)
plot(mapRK.upper, main = "90% Upper prediction limit",
breaks = brk, col = phCramp)
plot(mapRK.PIrange, main = "Prediction limit range",
col = terrain.colors(length(seq(0, 6.5, by = 1)) - 1), axes = FALSE,
breaks = seq(0, 6.5, by = 1))
```

7.3.3 Validating the Quantification of Uncertainty

For validation we assess both the quality of the predictions and the quantification of uncertainty as was done earlier for the universal kriging. Below we assess the validation of the cubist model alone and the cubist regression kriging model.

```
# Cubist model prediction
vPreds <- predict(hv.cub.Exp,  newdata = vDat)
# Residual prediction
coordinates(vDat) <- ~X + Y
OK.preds.V <- as.data.frame(krige(residual ~ 1, cDat, model = model_1,
newdata = vDat))

## [using ordinary kriging]

# Regression kriging predictions
OK.preds.V$cubist <- vPreds
OK.preds.V$finalP <- OK.preds.V$cubist + OK.preds.V$var1.pred

# Validation regression
goof(observed = vDat$pH60_100cm, predicted = OK.preds.V$cubist)

##          R2 concordance      MSE     RMSE      bias
## 1 0.2218322    0.422689 1.352345 1.162903 0.2017115

# regression kriging
goof(observed = vDat$pH60_100cm, predicted = OK.preds.V$finalP)

##          R2 concordance      MSE   RMSE      bias
## 1 0.2951672   0.5276429 1.259782 1.1224 0.1082966
```

The regression kriging model is a little more accurate than the Cubist model alone.

The first step in estimating the uncertainty about the validation points is to evaluate with rule set or partition it belongs to.

```
# Assign a rule to each observation if more than one obervation
vDat1 <- as.data.frame(vDat)
# Insert for rules bit
vDat1$rule = 9999
# rule 1
vDat1$rule[which(vDat1$NDVI <= -0.191489 & vDat1$TWI <= 13.17387)] <- 1
# rule 2
vDat1$rule[which(vDat1$TWI > 13.17387)] <- 2
## rule 3
vDat1$rule[which(vDat1$NDVI > -0.191489 & vDat1$TWI <= 13.17387)] <- 3

vDat1$rule <- as.factor(vDat1$rule)
summary(vDat1$rule)

##  1  2  3
## 43 65 44

# append regression kriging predictions
vDat1 <- cbind(vDat1, OK.preds.V[, "finalP"])
names(vDat1)[ncol(vDat1)] <- "RKpred"
```

Then we can define the prediction interval that corresponds to each observation for each level of confidence.

```
# Upper PL
ulMat <- matrix(NA, nrow = nrow(vDat1), ncol = length(r1.q))
for (i in seq(2, 20, 2)) {
    ulMat[which(vDat1$rule == 1), i] <- r1.q[i]
    ulMat[which(vDat1$rule == 2), i] <- r2.q[i]
    ulMat[which(vDat1$rule == 3), i] <- r3.q[i]
}

# Lower PL
for (i in seq(1, 20, 2)) {
    ulMat[which(vDat1$rule == 1), i] <- r1.q[i]
    ulMat[which(vDat1$rule == 2), i] <- r2.q[i]
    ulMat[which(vDat1$rule == 3), i] <- r3.q[i]
}

# upper and lower prediction limits
ULpreds <- ulMat + vDat1$RKpred

# binary
bMat <- matrix(NA, nrow = nrow(ULpreds), ncol = (ncol(ULpreds)/2))
cnt <- 1
for (i in seq(1, 20, 2)) {
    bMat[, cnt] <- as.numeric(vDat1$pH60_100cm <= ULpreds[, i + 1]
    & vDat1$pH60_100cm >= ULpreds[, i])
    cnt <- cnt + 1
}

colSums(bMat)/nrow(bMat)

##  [1] 0.99342105 0.98026316 0.94078947 0.88157895 0.78947368
         0.59210526
##  [7] 0.48026316 0.23026316 0.13157895 0.05921053
```

The PICP estimates appear to correspond quite well with the respective confidence levels. This can be observed from the plot on Fig. 7.6 too.

```
# make plot
cs <- c(99, 97.5, 95, 90, 80, 60, 40, 20, 10, 5)
# confidence level
plot(cs, ((colSums(bMat)/nrow(bMat)) * 100))
abline(a = 0, b = 1, lty = 2, col = "red")
```

From the validation observations the prediction intervals range between 2.45 and 4.6 with a median of about 3.36 pH units when using the Cubist regression kriging model.

```
quantile(ULpreds[, 8] - ULpreds[, 7])

##       0%      25%      50%      75%     100%
## 2.456651 2.456651 3.361965 4.603535 4.603535
```

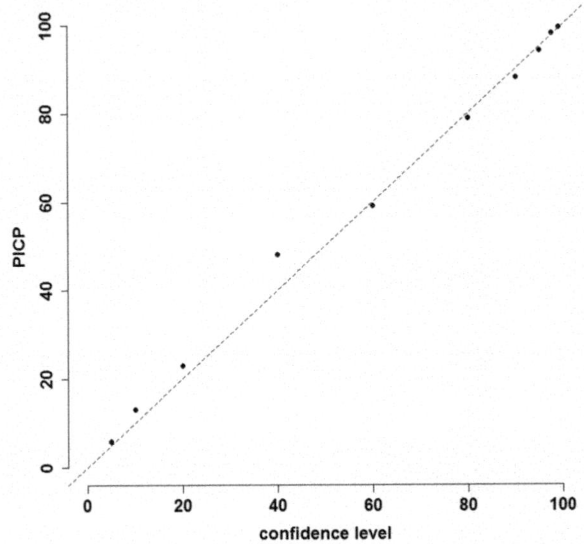

Fig. 7.6 Plot of PICP and confidence level based on validation of Cubist regression kriging model

7.4 Empirical Uncertainty Quantification Through Fuzzy Clustering and Cross Validation

Like the previous uncertainty approach, this approach is similar in that uncertainty is expressed in the form of quantiles of the underlying distribution of model error (residuals). It contrasts however in terms of how the environmental data is partitioned. For the previous approach, partitioning was based upon the hard classification defined by a fitted Cubist model. This approach uses a fuzzy clustering method of partition.

Essentially the approach is based on previous research by Shrestha and Solomatine (2006) where the idea is to partition a feature space into clusters (with a fuzzy k-means routine) which share similar model errors. A prediction interval (PI) is constructed for each cluster on the basis of the empirical distribution of residual observations that belong to each cluster. A PI is then formulated for each observation in the feature space according to the grade of their memberships to each cluster. They applied this methodology to artificial and real hydrological data sets and it was found to be superior to other methods which estimate a PI. The Shrestha and Solomatine (2006) approach computes the PI independently and while free of the prediction model structure, it requires only the model or prediction outputs. Tranter et al. (2010) extended this approach to deal with observations that are outside of the training domain.

The method presented in this exercise was introduced by Malone et al. (2011) which modifies slightly the Shrestha and Solomatine (2006) and Tranter et al. (2010) approaches to enable it for a DSM framework. The approach is summarized by the flow diagram on Fig. 7.7.

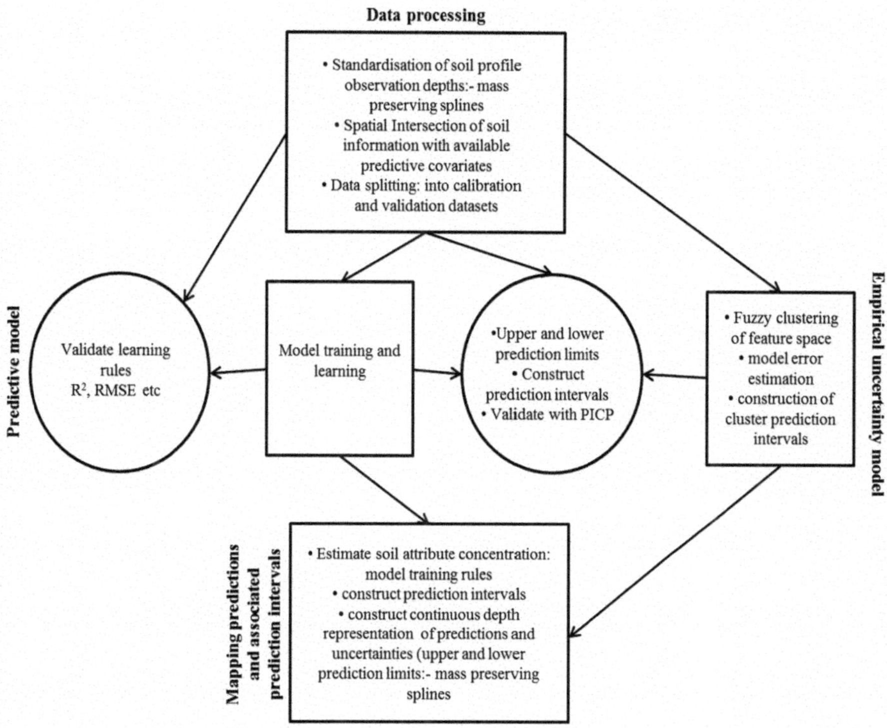

Fig. 7.7 Flow diagram of the general procedure for achieving the outcome of mapping predictions and their uncertainties (upper and lower prediction limits) within a digital soil mapping framework. The 3 components for achieving this outcome are the prediction model, the empirical uncertainty model and the mapping component (Sourced from Malone et al. 2011)

The process for deriving the uncertainties is much the same as for the previous approach using the Cubist regression kriging approach. One benefit of using a fuzzy kmeans approach is that the spatial distribution of uncertainty is represented as a continuous variable. Further, the incorporation of extragrades in the fuzzy kmeans classifying provides an explicit means to identify and highlight areas of the greatest uncertainty and possibly where new sampling efforts should be prioritized. As shown on Fig. 7.7 the approach entails 3 main processes:

- Calibrating the spatial model
- deriving the uncertainty model which includes both estimations of model errors and fuzzy kmeans with extragrades classification
- Creation of maps of both spatial soil predictions and uncertainties.

Naturally, this framework is validated using a withheld or better still independent data set.

7.4.1 Defining the Model Parameters

Here we will use a random forest regression kriging model for the prediction of soil pH across the study area. This model will also be incorporated into the uncertainty model via leave-one-out cross validation in order to derive the model errors. As before, we begin by preparing the data.

```
# Point data subset data for modeling
set.seed(667)
training <- sample(nrow(HV_subsoilpH), 0.7 * nrow(HV_subsoilpH))
cDat <- HV_subsoilpH[training, ]
vDat <- HV_subsoilpH[-training, ]
```

Now we fit a random forest model using all possible covariates.

```
# fit the model
library(randomForest)
hv.RF.Exp <- randomForest(pH60_100cm ~ Terrain_Ruggedness_Index
+ AACN + Landsat_Band1 + Elevation + Hillshading
+ Light_insolation + Mid_Slope_Positon + MRVBF + NDVI + TWI
+ Slope,data = cDat, importance = TRUE, ntree = 1000)

# Goodness of fit
goof(observed = cDat$pH60_100cm, predicted = predict(hv.RF.Exp,
   newdata = cDat), plot.it = FALSE)

##           R2 concordance       MSE      RMSE        bias
## 1 0.9311825   0.8993517 0.2718095 0.5213536 0.002527227
```

The hv.RF.Exp model appears to perform quite well when we examine the goodness of fit statistics.

Now we can examine the model residuals for any presence of spatial structure with variogram modeling. For the output below it does seems that there is some useful correlation structure in the residuals that will likely help to improve upon the performance of the hv.RF.Exp model.

```
# Estimate the residual
cDat$residual <- cDat$pH60_100cm - predict(hv.RF.Exp,
newdata = cDat)
# residual variogram model
coordinates(cDat) <- ~X + Y
vgm1 <- variogram(residual ~ 1, data = cDat, width = 200)
mod <- vgm(psill = var(cDat$residual), "Sph", range = 10000,
   nugget = 0)
model_1 <- fit.variogram(vgm1, mod)
model_1

##   model      psill    range
## 1   Nug 0.18822373    0.000
## 2   Sph 0.09624309 1307.933
```

```
gOK <- gstat(NULL, "hunterpH_residual_RF", residual ~ 1, cDat,
   model = model_1)
```

Like before, we need to estimate the model errors and a good way to do this is via a LOCV approach. The script below is more-or-less a repeat from earlier with the Cubist regression kriging modeling except now we are using the random forest modeling.

```
# Uncertainty analysis
cDat1 <- as.data.frame(cDat)
names(cDat1)
cDat1.r1 <- cDat1
target.C.r1 <- cDat1.r1$pH60_100cm

looResiduals <- numeric(nrow(cDat1.r1))
for (i in 1:nrow(cDat1.r1)) {
    looRFPred <- randomForest(pH60_100cm ~ Terrain_Ruggedness_Index
    + AACN + Landsat_Band1 + Elevation + Hillshading
    + Light_insolation + Mid_Slope_Positon + MRVBF + NDVI
    + TWI + Slope,
    data = cDat1.r1[-i, ], importance = TRUE,
    ntree = 1000)

    cDat11.r1.sub <- cDat1.r1[-i, ]
    cDat11.r1.sub$pred <- predict(looRFPred, newdata = cDat11.r1.sub)
    cDat11.r1.sub$resids <- target.C.r1[-i] - cDat11.r1.sub$pred
    # residual variogram
    vgm.r1 <- variogram(resids ~ 1, ~X + Y, cDat11.r1.sub,
    width = 200)
    mod.r1 <- vgm(psill = var(cDat11.r1.sub$resids), "Sph",
    range = 10000,
    nugget = 0)
    model_1.r1 <- fit.variogram(vgm.r1, mod.r1)
    model_1.r1
    # interpolate residual
    int.resids1.r1 <- krige(cDat11.r1.sub$resids ~ 1,
    locations = ~X + Y,
    data = cDat11.r1.sub, newdata = cDat1.r1[i, c("X", "Y")], model
    = model_1.r1, debug.level = 0)[, 3]
    looPred <- predict(looRFPred, newdata = cDat1.r1[i, ])
    looResiduals[i] <- target.C.r1[i] - (looPred + int.resids1.r1)
}

# Combine residual to main data frame
cDat1 <- cbind(cDat1, looResiduals)
```

```
##  [1] "X"                         "Y"
##  [3] "pH60_100cm"                "Terrain_Ruggedness_Index"
##  [5] "AACN"                      "Landsat_Band1"
##  [7] "Elevation"                 "Hillshading"
##  [9] "Light_insolation"          "Mid_Slope_Positon"
```

```
## [11] "MRVBF"                        "NDVI"
## [13] "TWI"                          "Slope"
## [15] "residual"                     "looResiduals"
```

The defining aspect of this uncertainty approach is fuzzy clustering with extragrades. McBratney and de Gruijter (1992) recognized a limitation of the normal fuzzy clustering algorithm in that it had an inability to distinguish between observations very far from the cluster centroids and those at the centre of the centroid configuration. These observations were termed extragrades as opposed to intragrades, which are the observations that lie between the main clusters. The extragrades are considered the outliers of the data set and have a distorting influence on the configuration of the main clusters (Lagacherie et al. 1997). McBratney and de Gruijter (1992) developed an adaptation to the FKM algorithm which distinguishes observations that should belong to an extragrade cluster. The FKM with extragrades algorithm minimizes the objective function:

$$J_e(\mathbf{C},\mathbf{M}) = \alpha \sum_{i=1}^{n}\sum_{j=1}^{c} m_{ij}^{\psi} d_{ij}^{2} + (1-\alpha)\sum_{i=1}^{n} m_{i\star}^{\psi} \sum_{i=1}^{n} d_{ij}^{-}2 \tag{7.2}$$

where $\mathbf{C}$ is the $c \times p$ matrix of cluster centers where c is the cluster and p is the number of variables. $\mathbf{M}$ is the $n \times c$ matrix of partial memberships, where n is the number of observations; $m_{ij} \epsilon [0, 1]$ is the partial membership of the ith observation to the jth cluster. $\psi \geq 1$ is the fuzziness exponent. The square distance between the ith observation to the jth cluster is d_{ij}^{2}. $m_{i\star}^{\psi}$ denotes the membership to the extragrade cluster. This function also requires the parameter alpha (α) to be defined, which is used to evaluate membership to the extragrade cluster.

A very good stand-alone software developed by Minasny and McBratney (2002a) called 'Fuzme' contains the FKM with extragrades method, plus other clustering methods. The software may be downloaded for free from http://sydney.edu.au/agriculture/pal/software/fuzme.shtml. The source script to this software has also been written to an R package of the same name. Normally, the stand-alone software would be used because it is computationally much faster. However, using the fuzme R package allows one to easily integrate the clustering procedures into a standard R workflow. For example, one of the issues of clustering procedures is the uncertainty regarding the selection of an optimal cluster number for a given data set. There are a number of ways to determine an optimal solution. Some popular approaches include to determining the cluster combination which minimizes the Fuzzy Performance Index (FPI) or Modified partition Entropy (MPE). The MPE establishes the degree of fuzziness created by a specified number of clusters for a defined exponent value. The notion is that the smaller the MPE, the more suitable is the corresponding number of clusters at the given exponent value. A more sophisticated analysis is to look at the derivative of $J_e(\mathbf{C},\mathbf{M})$ with respect to ψ and is used to simultaneously establish the optimal ψ and cluster size. More is discussed about each of these indices by Odeh et al. (1992) and Bragato (2004).

The key generally is to define clusters which can be easily distinguished compared to a collection of clusters that are all quite similar. Such diagnostic criteria are useful; however it would be more beneficial to determine an optimal cluster configuration based on criteria that are meaningful to the current situation of deriving prediction uncertainties. In this case we might want to evaluate the optimal cluster number *and* fuzzy exponent that maximizes the fidelity of PICP to confidence interval, and minimizes the prediction interval range. Hence in this section we will integrate the `R` fuzme with the DSM uncertainties workflow to achieve those ends.

The idea here is to determine an optimal cluster number and fuzzy exponent based on criteria related to prediction interval width and PICP, together with some of the other fuzzy clustering criteria. We essentially want to perform fuzzy clustering with extragrades upon the environmental covariate of the data points in the `cdat1` object. Fuzzy clustering with extragrades is implemented via a two-step procedure using the `fobjk` and `fkme` functions from the fuzme package. The `fobjk` function allows one to find an optimal solution for alpha (α) which is the parameter that allows one to control the membership of data to the extragrade cluster. For example, if we wanted to stipulate that we want the average extragrade membership of a data set to be 10 %, then we need to find an optimal solution for alpha to achieve that outcome. This can be controlled through the `Uereq` parameter of the `fobjk` function.

The first step is to load the fuzme library and prepare the input data.

```
library(fuzme) install_bitbucket("brendo1001/fuzme/rPackage/
   fuzme")
library(devtools)
```

Now we need to prepare the data for clustering, and parameterize the other inputs of the function. The other inputs are: `nclass`, which is the number of clusters you want to define. Note that an extra cluster will be defined as this will be the extragrade cluster and the associated memberships. `data` is the data needed for clustering, `U` is an initial membership matrix in order to get the fuzzy clustering algorithm operable. `phi` is the fuzzy exponent, while `distype` refers to the distance metric to be used for clustering. There are 3 possible distance metrics available. These are: Euclidean (1), Diagonal (2), and Mahalanobis (3). As an example of using the `fobjk` lets define 4 clusters with a fuzzy exponent of 1.2, and with the average extragrade membership of 10 %. Currently this function is pretty slow to compute the optimal alpha, so be prepared to wait a while.

```
# Parameterize fuzzy objective function
data.t <- cDat1[, 4:14]  # data to be clustered
nclass.t <- 4  # number of clusters
phi.t <- 1.2
distype.t <- 3  #3 = Mahalanobis distance
Uereq.t <- 0.1  #average extragrade membership

# initial membership matrix
```

```
scatter.t <- 0.2  # scatter around initial membership
ndata.t <- dim(data.t)[1]
U.t <- initmember(scatter = scatter.t, nclass = nclass.t,
  ndata = ndata.t)

# run fuzzy objective function
alfa.t <- fobjk(Uereq = Uereq.t, nclass = nclass.t, data = data.t,
   U = U.t, phi = phi.t, distype = distype.t)
```

Remember the `fobjk` function will only return the optimal alfa value. This value then gets inserted into the associated `fkme` function in order to estimate the memberships of the data to each cluster and the extragrade cluster. The `fkme` function also returns the cluster centroids too.

```
alfa.t

##          1
## 0.01136809
```

The `fkme` function is parameterized similarly to `fobjk`, yet with some additional inputs related to the number of allowable iterations for convergence (`maxiter`), the convergence criterion value (`toldif`), and whether or not to display behind-the-scenes processing.

```
tester <- fkme(nclass = nclass.t, data = data.t, U = U.t, phi = phi.t,
alfa = alfa.t, maxiter = 500, distype = distype.t, toldif = 0.01,
verbose = 1)
```

The `fkme` function returns a list with a number of elements. At this stage we are primarily interested in the elements `membership` and `centroid` which we will use a little later on.

As described earlier, there are a number of criteria to assess the validity of a particular clustering configuration. We can evaluate these by using the `fvalidity` function. It essentially takes in a few of the outputs from the `fkme` function.

```
fvalidity(U = tester$membership[, 1:4], dist = tester$distance,
centroid = tester$ centroid, nclass = 4, phi = 1.2,
W = tester$distNormMat)

##         fpi       mpe         S  dJ/dphi
## 1 0.5149733 0.4026764 0.8377058 -1710.19
```

Another useful metric is the confusion index (after Burrough et al. 1997) which in our case looks at the similarity between the highest and second highest cluster memberships. The confusion index is estimated for each data point. Taking the average over the data set provides some sense of whether cluster can be distinguished from each other.

```
mean(confusion(nclass = 5, U = tester$membership))

## [1] 0.3356157

# Note number of cluster is set to 5 to account for the additional
# extragrade cluster.
```

To assess the clustering performance using the criteria of the PICP and prediction interval range, we need to first assign each data point a one of the clusters we have derived. The assignment is based on the cluster which has the highest membership grade. The script below provides a method for evaluating which data point belongs to which cluster

```
membs <- as.data.frame(tester$membership)
membs$class <- 99999
for (i in 1:nrow(membs)) {
    mbs2 <- membs[i, 1:ncol(tester$membership)]
    # which is the maximum probability on this row
    membs$class[i] <- which(mbs2 == max(mbs2))[1]
}
membs$class <- as.factor(membs$class)
summary(membs$class)

##  1  2  3  4  5
## 58 84 78 90 44
```

Then we combine it to the `cDat1` object which contains the information regarding the model errors (specifically in the `looResiduals` column).

```
# combine
cDat1 <- cbind(cDat1, membs$class)
names(cDat1)[ncol(cDat1)] <- "class"
levels(cDat1$class)

## [1] "1" "2" "3" "4" "5"
```

Then we derive the cluster model error. This entails splitting the `cDat1` object up on the basis of the cluster with the highest membership i.e. `cDat1$class`.

```
cDat2 <- split(cDat1, cDat1$class)

# cluster lower prediction limits
quanMat1 <- matrix(NA, ncol = 10, nrow = length(cDat2))
for (i in 1:length(cDat2)) {
    quanMat1[i, ] <- quantile(cDat2[[i]][, "looResiduals"],
    probs = c(0.005, 0.0125, 0.025, 0.05, 0.1, 0.2, 0.3, 0.4,
    0.45, 0.475), na.rm = FALSE, names = F, type = 7)
}

row.names(quanMat1) <- levels(cDat1$class)
quanMat1[nrow(quanMat1), ] <- quanMat1[nrow(quanMat1), ] * 2
```

```
quanMat1 <- t(quanMat1)
row.names(quanMat1) <- c(99, 97.5, 95, 90, 80, 60, 40, 20, 10, 5)#

# cluster upper prediction limits
quanMat2 <- matrix(NA, ncol = 10, nrow = length(cDat2))
for (i in 1:length(cDat2)) {
    quanMat2[i, ] <- quantile(cDat2[[i]][, "looResiduals"],
    probs = c(0.995, 0.9875, 0.975, 0.95, 0.9, 0.8, 0.7, 0.6,
    0.55, 0.525), na.rm = FALSE, names = F, type = 7)
}

quanMat2[quanMat2 < 0] <- 0
row.names(quanMat2) <- levels(cDat1$class)
quanMat2[nrow(quanMat2), ] <- quanMat2[nrow(quanMat2), ] * 2
quanMat2 <- t(quanMat2)
row.names(quanMat2) <- c(99, 97.5, 95, 90, 80, 60, 40, 20, 10, 5)#
```

The objects quanMat1 and quanMat2 represent the lower and upper model errors for each cluster for each quantile respectively. For the extragrade cluster, we multiple the error by a constant, here 2, in order to explicitly indicate that the extragrade cluster (being outliers of the data) have a higher uncertainty.

Using the validation or independent data that has been withheld, we evaluate the PICP and prediction interval width. This requires first allocating cluster memberships to the points on the basis of outputs from using the fkme function, then using these together with the cluster prediction limits to evaluate weighted averages of the prediction limits for each point. With that done we can then derive the unique upper and lower prediction interval limits for each point at each confidence level. First, for the membership allocation, we use the fuzExall function. Essentially this function takes in outputs from the fkme function and in our case, specifically that concerning the tester object. Recall that the validation data is saved to the vDat object.

```
vDat1 <- as.data.frame(vDat)
names(vDat1)

##  [1] "X"                    "Y"
##  [3] "pH60_100cm"           "Terrain_Ruggedness_Index"
##  [5] "AACN"                 "Landsat_Band1"
##  [7] "Elevation"            "Hillshading"
##  [9] "Light_insolation"     "Mid_Slope_Positon"
## [11] "MRVBF"                "NDVI"
## [13] "TWI"                  "Slope"

# covariates of the validation data
vCovs <- vDat1[, c("Terrain_Ruggedness_Index", "AACN",
"Landsat_Band1", "Elevation", "Hillshading", "Light_insolation",
"Mid_Slope_Positon", "MRVBF", "NDVI", "TWI", "Slope")]

# run fkme allocation function
fuzAll <- fuzExall(data = vCovs, phi = 1.2, centroid
```

```
= tester$centroid, distype = 3, W = tester$distNormMat,
alfa = tester$alfa)

# Get the memberships
fuz.me <- fuzAll$membership
```

A "fuzzy committee" approach is used to estimate the underlying residual at each point. In this case the upper and lower limits are derived by weighted mean of the lower and upper model errors of each cluster, where the weights are the cluster memberships. This can be defined mathematically as:

$$PI_i^L = \sum_{j=1}^{c} m_{ij} PIC_j^L \text{ and } PI_i^U = \sum_{j=1}^{c} m_{ij} PIC_j^U \tag{7.3}$$

where PI_i^L and PI_i^U correspond to the weighted lower and upper limits for the ith observation. PIC_j^L and PIC_j^U are the lower and upper limits for each cluster j, and m_{ij} is the membership grade of the ith observation to cluster j (which were derived in the previous step). In R, this can be interpreted as:

```
# lower prediction limit
lPI.mat <- matrix(NA, nrow = nrow(fuz.me), ncol = 10)
for (i in 1:nrow(lPI.mat)) {
    for (j in 1:nrow(quanMat1)) {
        lPI.mat[i, j] <- sum(fuz.me[i, 1:ncol(fuz.me)]
            * quanMat1[j, ])
    }
}

# upper prediction limit
uPI.mat <- matrix(NA, nrow = nrow(fuz.me), ncol = 10)
for (i in 1:nrow(uPI.mat)) {
    for (j in 1:nrow(quanMat2)) {
        uPI.mat[i, j] <- sum(fuz.me[i, 1:ncol(fuz.me)]
            * quanMat2[j, ])
    }
}
```

Then we want to add these values to the actual regression kriging predictions.

```
# Regression kriging predictions
vPreds <- predict(hv.RF.Exp, newdata = vDat)
coordinates(vDat) <- ~X + Y
OK.preds.V <- as.data.frame(krige(residual ~ 1, cDat,
model = model_1, newdata = vDat))

## [using ordinary kriging]

OK.preds.V$randomForest <- vPreds
OK.preds.V$finalP <- OK.preds.V$randomForest + OK.preds.V$var1.pred
```

```
# Add prediction limits to regression kriging predictions
vDat1 <- cbind(vDat1, OK.preds.V$finalP)
names(vDat1)[ncol(vDat1)] <- "RF_rkFin"

# Derive validation lower prediction limits
lPL.mat <- matrix(NA, nrow = nrow(fuz.me), ncol = 10)
for (i in 1:ncol(lPL.mat)) {
    lPL.mat[, i] <- vDat1$RF_rkFin + lPI.mat[, i]
}

# Derive validation upper prediction limits
uPL.mat <- matrix(NA, nrow = nrow(fuz.me), ncol = 10)
for (i in 1:ncol(uPL.mat)) {
    uPL.mat[, i] <- vDat1$RF_rkFin + uPI.mat[, i]
}
```

Now as in the previous uncertainty approaches we estimate the PICP for each level of confidence. We can also estimate the average prediction interval length too. We will do this for the 90 % confidence level.

```
bMat <- matrix(NA, nrow = nrow(fuz.me), ncol = 10)
for (i in 1:10) {
    bMat[, i] <- as.numeric(vDat1$pH60_100cm <= uPL.mat[, i]
    & vDat1$pH60_100cm >= lPL.mat[, i])
}

# PICP
colSums(bMat)/nrow(bMat)

##  [1] 0.9736842 0.9407895 0.9144737 0.8815789 0.7960526
        0.5723684 0.3881579
##  [8] 0.2039474 0.1250000 0.1118421

# prediction interval width (averaged)
as.matrix(mean(uPL.mat[, 4] - lPL.mat[, 4]))

##          [,1]
## [1,] 3.913579
```

Recall that our motivation at the moment to to derive and optimal cluster number and fuzzy exponent based on criteria of the PICP and prediction interval width. Above were the steps for evaluating those values for one clustering parameter configuration i.e. 4 clusters (plus and extragrade) with a fuzzy exponent of 1.2. Essentially we need to run sequentially, different combinations of cluster number and fuzzy exponent value and then assess to criteria resulting from each of the different configurations in order to find the optimum. For example we might initiate the process by specifying:

```
nclass.t <- seq(2, 6, 1)
nclass.t

## [1] 2 3 4 5 6

phi.t <- seq(1.2, 1.6, 0.1)
phi.t

## [1] 1.2 1.3 1.4 1.5 1.6
```

Then it is a matter of using all possible combinations of these values in the `fobjk` and `fkme` functions that are needed ultimately to estimate the PICP and prediction interval width as was just done above for a single combination. Computationally this can be pretty complex to organize. However, for brevity, below are the clustering outputs from implementing the described procedure for finding the optimal clustering parameter configuration.

```
fkme.outs

##    class expon        alfa confusion       fpi       mpe            S
## 1      2   1.2 0.005124426 0.3826885 0.5664630 0.6686026 2.434422e+00
## 2      2   1.3 0.005379610 0.5546698 0.7839436 0.8995941 3.054369e+00
## 3      2   1.4 0.005247453 0.6992748 0.9354939 1.0541716 4.641781e+00
## 4      2   1.5 0.004997139 0.8418018 1.0441222 1.1612614 1.396096e+01
## 5      2   1.6 0.004700295 0.9707954 1.0915822 1.2156737 1.466621e+06
## 6      3   1.2 0.008058919 0.3724314 0.4757238 0.5303243 1.981344e+00
## 7      3   1.3 0.008700468 0.5496682 0.6911314 0.7563128 2.668302e+00
## 8      3   1.4 0.009085534 0.7501051 0.8629492 0.9272163 1.732949e+01
## 9      3   1.5 0.009135423 0.8499948 0.9400907 1.0114920 3.759586e+09
## 10     3   1.6 0.008965663 0.9622403 1.0269282 1.0983380 1.931253e+06
## 11     4   1.2 0.011368089 0.3356154 0.4098895 0.4404190 1.466233e+00
## 12     4   1.3 0.012443913 0.5194940 0.6338683 0.6721924 1.612367e+00
## 13     4   1.4 0.013312052 0.7052652 0.8180113 0.8575417 3.673102e+00
## 14     4   1.5 0.013979655 0.8456542 0.9112886 0.9580570 3.265964e+08
## 15     4   1.6 0.014225644 0.9552375 1.0052042 1.0566683 1.238797e+07
## 16     5   1.2 0.014470766 0.3284219 0.3786159 0.3889102 1.346230e+00
## 17     5   1.3 0.016338497 0.5085432 0.6032573 0.6227569 1.345716e+00
## 18     5   1.4 0.018119534 0.7390841 0.8163015 0.8378098 3.718401e+04
## 19     5   1.5 0.019640328 0.8469274 0.9033973 0.9361176 3.123211e+07
## 20     5   1.6 0.020083903 0.9506230 0.9945531 1.0345826 1.146549e+07
## 21     6   1.2 0.018138555 0.3353890 0.3670481 0.3605114 1.231055e+00
## 22     6   1.3 0.020091380 0.4990245 0.5730767 0.5777931 1.222209e+00
## 23     6   1.4 0.023620259 0.7438648 0.8203979 0.8315384 2.894973e+05
## 24     6   1.5 0.025368293 0.8495561 0.9027086 0.9267227 1.174473e+11
## 25     6   1.6 0.026754212 0.9462337 0.9880089 1.0205813 1.641539e+09
##       dJdphi       PICP      PIw
## 1  -1156.906 0.12973684 3.975894
## 2  -1434.856 0.15605263 3.992315
## 3  -1558.146 0.14421053 3.885237
## 4  -1594.820 0.17052632 3.883247
## 5  -1528.474 0.15736842 3.857267
## 6  -1361.192 0.11105263 3.861995
## 7  -1734.959 0.09789474 3.968063
## 8  -1922.302 0.14289474 3.976594
## 9  -1857.203 0.13368421 3.895518
```

```
## 10 -1803.114 0.33631579 4.483795
## 11 -1356.265 0.15184211 3.913579
## 12 -1810.981 0.17026316 3.995139
## 13 -2039.601 0.12947368 4.019338
## 14 -1973.390 0.14684211 4.197816
## 15 -1882.285 0.20473684 4.257101
## 16 -1351.156 0.21631579 3.903318
## 17 -1849.211 0.20052632 3.850294
## 18 -2161.102 0.15315789 3.833518
## 19 -2038.082 0.11000000 4.185526
## 20 -1897.424 0.29026316 4.499928
## 21 -1374.898 0.24263158 3.818262
## 22 -1833.395 0.20315789 3.848154
## 23 -2250.853 0.20052632 3.653220
## 24 -2075.234 0.13631579 4.155074
## 25 -1886.003 0.34947368 4.619364
```

The data frame above lists the optimal alfa, clustering validity diagnostics, PICP and prediction interval width diagnostics for each combination. The PICP column is actually an absolute distance of the PICP at each level of prescribed confidence. Subsequently we should look for a minimum in that regard. Overall the best combination considering PICP and PIw together is 3 clusters with a fuzzy exponent of either 1.2 or 1.3. Based on the other fuzzy validity criteria a fuzzy exponent of 1.2 is optimal. Now we just re-run the `function` with these optimal values in order to derive the cluster centroids which are need in order to create maps of the prediction interval and range.

```
U.t <- initmember(scatter = 0.2, nclass = 3, ndata = ndata.t)
fkme.fin <- fkme(nclass = 3, data = data.t, U = U.t, phi = 1.2,
alfa = 0.008058919, maxiter = 500, distype = 3, toldif = 0.01, verbose = 1)
```

Now we have to calculate the cluster model error limits. This is achieved by evaluating which cluster each data point in `cDat1` belongs to based on the maximum membership. Then we derive the quantiles of the model errors for each cluster.

```
# Assign cluster to respective data point
membs <- as.data.frame(fkme.fin$membership)
membs$class <- 99999
for (i in 1:nrow(membs)) {
    mbs2 <- membs[i, 1:ncol(fkme.fin$membership)]
    # which is the maximum probability on this row
    membs$class[i] <- which(mbs2 == max(mbs2))[1]
}
membs$class <- as.factor(membs$class)
summary(membs$class)
```

```
##   1   2   3   4
## 119 104  91  40
```

```
# combine to main data frame
cDat1 <- cbind(cDat1, membs$class)
names(cDat1)[ncol(cDat1)] <- "class"
levels(cDat1$class)

## [1] "1" "2" "3" "4"

# split data frame based on class
cDat2 <- split(cDat1, cDat1$class)

# cluster lower prediction limits
quanMat1 <- matrix(NA, ncol = 10, nrow = length(cDat2))
for (i in 1:length(cDat2)) {
    quanMat1[i, ] <- quantile(cDat2[[i]][, "looResiduals"],
    probs = c(0.005, 0.0125, 0.025, 0.05, 0.1, 0.2, 0.3, 0.4,
    0.45, 0.475), na.rm = FALSE, names = F, type = 7)
}

row.names(quanMat1) <- levels(cDat1$class)
quanMat1[nrow(quanMat1), ] <- quanMat1[nrow(quanMat1), ] * 2
quanMat1 <- t(quanMat1)
row.names(quanMat1) <- c(99, 97.5, 95, 90, 80, 60, 40, 20, 10, 5)

# cluster upper prediction limits
quanMat2 <- matrix(NA, ncol = 10, nrow = length(cDat2))
for (i in 1:length(cDat2)) {
    quanMat2[i, ] <- quantile(cDat2[[i]][, "looResiduals"],
    probs = c(0.995, 0.9875, 0.975, 0.95, 0.9, 0.8, 0.7, 0.6,
    0.55, 0.525), na.rm = FALSE, names = F, type = 7)
}

quanMat2[quanMat2 < 0] <- 0
row.names(quanMat2) <- levels(cDat1$class)
quanMat2[nrow(quanMat2), ] <- quanMat2[nrow(quanMat2), ] * 2
quanMat2 <- t(quanMat2)
row.names(quanMat2) <- c(99, 97.5, 95, 90, 80, 60, 40, 20, 10, 5)
```

7.4.2 Spatial Mapping

With the spatial model defined together with the fuzzy clustering with the associated parameters and model errors, we can create the associated maps. First, the random forest regression kriging map. The map is shown on Fig. 7.9.

```
map.Rf <- predict(hunterCovariates_sub, hv.RF.Exp,
filename = "RF_HV.tif", format = "GTiff", overwrite = T)
```

```
# kriged residuals
crs(hunterCovariates_sub) <- NULL
map.KR <- interpolate(hunterCovariates_sub, gOK, xyOnly = TRUE,
index = 1, filename = "krigedResid_RF.tif", format = "GTiff",
datatype = "FLT4S", overwrite = TRUE)

# raster stack of predictions and residuals
r2 <- stack(map.Rf, map.KR)
f1 <- function(x) calc(x, sum)

# add both maps
mapRF.fin <- calc(r2, fun = sum, filename = "RF_RF.tif",
format = "GTiff",
overwrite = T)
```

Now we need to map the prediction intervals. Essentially for every pixel on the map we first need to estimate the membership value to each cluster. This membership is based on a distance of the covariate space and the centroids of each cluster. To do this we use the fuzzy allocation function (`fuzExall`) that was used earlier. This time we use the fuzzy parameters from the `fkme_final` object. We need to firstly create a `dataframe` from the `rasterStack` of covariates.

```
# Prediction Intervals
hunterCovs.df <- data.frame(cellNos = seq(1:ncell(hunterCovariates_sub)))
vals <- as.data.frame(getValues(hunterCovariates_sub))
hunterCovs.df <- cbind(hunterCovs.df, vals)
hunterCovs.df <- hunterCovs.df[complete.cases(hunterCovs.df), ]
cellNos <- c(hunterCovs.df$cellNos)
gXY <- data.frame(xyFromCell(hunterCovariates_sub, cellNos, spatial = FALSE))
hunterCovs.df <- cbind(gXY, hunterCovs.df)
str(hunterCovs.df)

## 'data.frame': 33252 obs. of  14 variables:
##  $ x                       : num  340935 340960 340985 341010 341035 ...
##  $ y                       : num  6370416 6370416 6370416 6370416 ...
##  $ cellNos                 : int  101 102 103 104 105 106 107 108 ...
##  $ Terrain_Ruggedness_Index: num  0.745 0.632 0.535 0.472 0.486 ...
##  $ AACN                    : num  9.78 9.86 10.04 10.27 10.53 ...
##  $ Landsat_Band1           : num  68 63 59 62 56 54 59 62 54 56 ...
##  $ Elevation               : num  103 103 102 102 102 ...
##  $ Hillshading             : num  0.94 0.572 0.491 0.515 0.568 ...
##  $ Light_insolation        : num  1712 1706 1701 1699 1697 ...
##  $ Mid_Slope_Positon       : num  0.389 0.387 0.386 0.386 0.386 ...
##  $ MRVBF                   : num  0.376 0.765 1.092 1.54 1.625 ...
##  $ NDVI                    : num  -0.178 -0.18 -0.164 -0.169 -0.172 ...
##  $ TWI                     : num  16.9 17.2 17.2 17.2 17.2 ...
##  $ Slope                   : num  0.968 0.588 0.503 0.527 0.581 ...
```

Now we prepare all the other inputs for the `fuzExall` function, and then run it. This may take a little time.

```
# run fuzme allocation function
fuz.me_ALL <- fuzExall(data = hunterCovs.df[, 4:ncol(hunterCovs.df)],
centroid = fkme.fin$centroid, phi = 1.2, distype = 3,
W = fkme.fin$distNormMat, alfa = fkme.fin$alfa)
head(fuz.me_ALL$membership)
```

```
##           [,1]       [,2]       [,3]         [,4]
## [1,] 0.4198477 0.03991937 0.04619495 0.4940379882
## [2,] 0.8235166 0.04977852 0.07423964 0.0524652482
## [3,] 0.8701595 0.04153527 0.08148581 0.0068194501
## [4,] 0.8533413 0.03511594 0.10930323 0.0022395558
## [5,] 0.8360081 0.03625381 0.12709744 0.0006406516
## [6,] 0.8362873 0.03617278 0.12718138 0.0003585568
```

With the memberships estimated, lets visualize them by creating the associated membership maps (Fig. 7.8).

```
# combine
hvCovs <- cbind(hunterCovs.df[, 1:2], fuz.me_ALL)
# Create raster
map.class1mem <- rasterFromXYZ(hvCovs[, c(1, 2, 3)])
names(map.class1mem) <- "class_1"
map.class2mem <- rasterFromXYZ(hvCovs[, c(1, 2, 4)])
names(map.class2mem) <- "class_2"
map.class3mem <- rasterFromXYZ(hvCovs[, c(1, 2, 5)])
names(map.class3mem) <- "class_3"
```

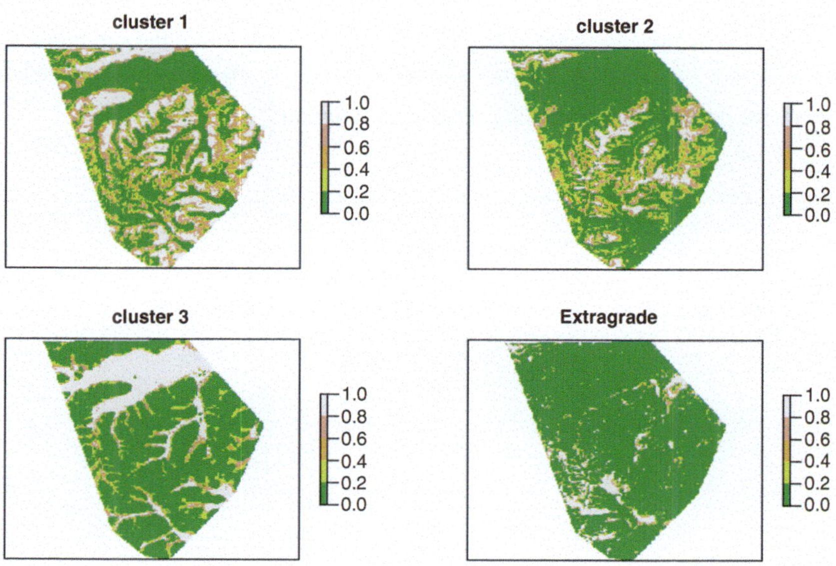

Fig. 7.8 Fuzzy cluster memberships

```
map.classExtramem <- rasterFromXYZ(hvCovs[, c(1, 2, 6)])
names(map.classExtramem) <- "class_ext"


par(mfrow = c(2, 2))
plot(map.class1mem, main = "cluster 1", col = terrain.colors
  (length(seq(0, 1, by = 0.2)) - 1), axes = FALSE, breaks =
   seq(0, 1, by = 0.2))
plot(map.class2mem, main = "cluster 2", col = terrain.colors
  (length(seq(0, 1, by = 0.2)) - 1), axes = FALSE, breaks =
   seq(0, 1, by = 0.2))
plot(map.class3mem, main = "cluster 3", col = terrain.colors
  (length(seq(0, 1, by = 0.2)) - 1), axes = FALSE, breaks =
   seq(0, 1, by = 0.2))
plot(map.classExtramem, main = "Extragrade", col = terrain.colors
   (length(seq(0, 1, by = 0.2)) - 1), axes = FALSE, breaks =
    seq(0, 1, by = 0.2))
```

The last spatial mapping task is to evaluate the 90 % prediction intervals. Again we use the fuzzy committee approach given the cluster memberships and the cluster model error limits.

```
# Lower limits
quanMat1["90", ]

##         1         2         3         4
## -1.532867 -1.451595 -1.792331 -3.140459

# upper limits
quanMat2["90", ]

##        1        2        3        4
## 2.023297 1.734574 2.182650 3.275685
```

Now we perform the weighted averaging prediction.

```
# Raster stack
s2 <- stack(map.class1mem, map.class2mem, map.class3mem,
    map.classExtramem)

# lower limit
f1 <- function(x) ((x[1] * quanMat1["90", 1]) + (x[2] *
quanMat1["90", 2]) + (x[3] * quanMat1["90", 3]) +
(x[4] * quanMat1["90", 4]))

mapRK.lower <- calc(s2, fun = f1, filename = "RF_lowerLimit.tif",
format = "GTiff", overwrite = T)
```

```
# upper limit
f1 <- function(x) ((x[1] * quanMat2["90", 1]) + (x[2] *
quanMat2["90", 2]) + (x[3] * quanMat2["90", 3]) + (x[4] *
quanMat2["90", 4]))

mapRK.upper <- calc(s2, fun = f1, filename = "RF_upperLimit.tif",
format = "GTiff", overwrite = T)
```

And finally we can derive the upper and lower prediction limits.

```
# raster stack
s3 <- stack(mapRF.fin, mapRK.lower, mapRK.upper)

# Lower prediction limit
f1 <- function(x) (x[1] + x[2])
mapRF.lowerPI <- calc(s3, fun = f1, filename = "RF_lowerPL.tif",
format = "GTiff", overwrite = T)

# Upper prediction limit
f1 <- function(x) (x[1] + x[3])
mapRF.upperPI <- calc(s3, fun = f1, filename = "RF_upperPL.tif",
format = "GTiff", overwrite = T)

# Prediction interval range
r2 <- stack(mapRF.lowerPI, mapRF.upperPI)
mapRF.PIrange <- calc(r2, fun = diff, filename = "cubistRK_PIrange.tif",
format = "GTiff", overwrite = T)
```

And now we can display the necessary maps (Fig. 7.9).

```
# color ramp
phCramp <- c("#d53e4f", "#f46d43", "#fdae61", "#fee08b", "#ffffbf",
"#e6f598", "#abdda4", "#66c2a5", "#3288bd", "#5e4fa2",
"#542788", "#2d004b")
brk <- c(2:14)
par(mfrow = c(2, 2))
plot(mapRF.lowerPI, main = "90% Lower prediction limit",
breaks = brk, col = phCramp)
plot(mapRF.fin, main = "Prediction", breaks = brk, col = phCramp)
plot(mapRF.upperPI, main = "90% Upper prediction limit",
breaks = brk, col = phCramp)
plot(mapRF.PIrange, main = "Prediction limit range",
col = terrain.colors(length(seq(0, 6.5, by = 1)) - 1), axes = FALSE,
breaks = seq(0, 6.5, by = 1))
```

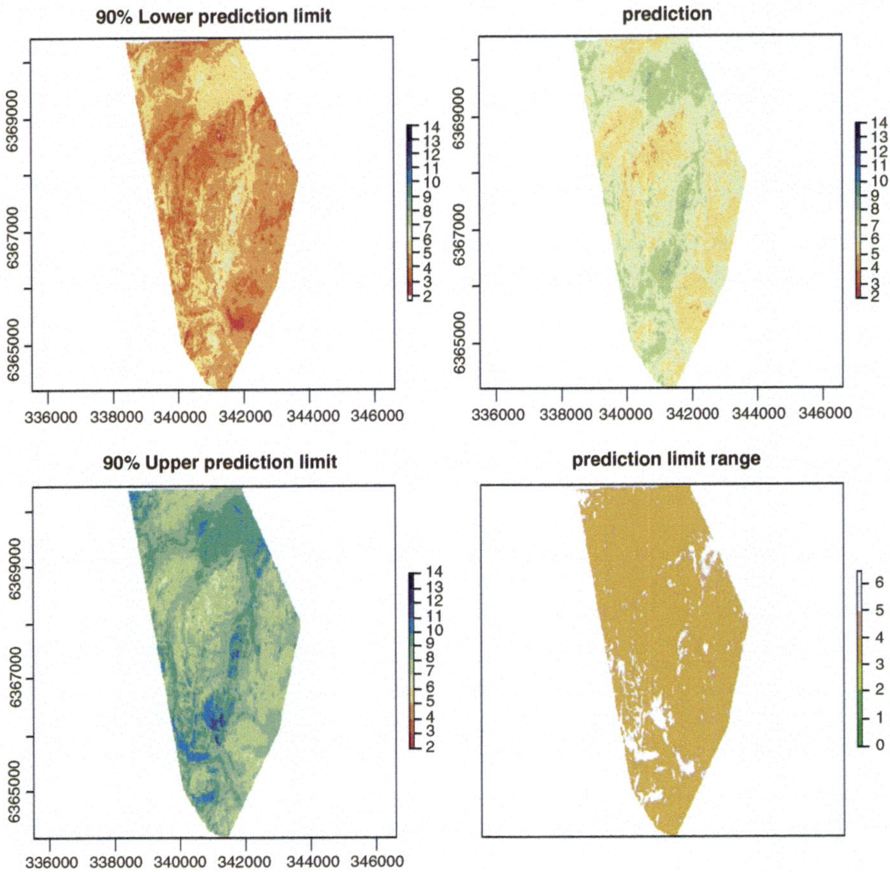

Fig. 7.9 Soil pH predictions and prediction limits derived using a Random Forest regression kriging prediction model together with LOCV and fuzzy classification

7.4.3 Validating the Quantification of Uncertainty

For the first step we can validate the random forest model alone and with the autocorrelated errors. As we had already applied the model earlier, it is just a matter of using the `goof` function to return the validation diagnostics.

```
# regression kriging
goof(observed = vDat$pH60_100cm, predicted = OK.preds.V$finalP)

##          R2 concordance      MSE   RMSE       bias
## 1 0.1904227   0.3733499 1.349082 1.1615 0.08998772
```

```
# Random Forest
goof(observed=vDat$pH60_100cm, predicted = OK.preds.V$randomForest)
```

```
##          R2 concordance      MSE     RMSE      bias
## 1 0.1055376   0.2592972 1.512683 1.229912 0.1306959
```

The regression kriging model performs better than the random forest model alone, but only marginally so; though both models are not particularly accurate in any case.

And now to validate the quantification of uncertainty we implement the workflow demonstrated above for the process of determining the optimal cluster parameter settings.

```
##  [1] "X"                       "Y"
##  [3] "pH60_100cm"              "Terrain_Ruggedness_Index"
##  [5] "AACN"                    "Landsat_Band1"
##  [7] "Elevation"               "Hillshading"
##  [9] "Light_insolation"        "Mid_Slope_Positon"
## [11] "MRVBF"                   "NDVI"
## [13] "TWI"                     "Slope"
## [using ordinary kriging]
```

Now lets evaluate the PICP for each level of confidence

```
bMat <- matrix(NA, nrow = nrow(fuz.me), ncol = 10)
for (i in 1:10) {
    bMat[, i] <- as.numeric(vDat1$pH60_100cm <= uPL.mat[, i] &
    vDat1$pH60_100cm >= lPL.mat[, i])
}

# PICP
colSums(bMat)/nrow(bMat)
```

```
##  [1] 0.9868421 0.9605263 0.9210526 0.8881579 0.7960526
         0.5723684 0.3947368
##  [8] 0.1842105 0.1250000 0.1184211
```

Plotting the PICP against the confidence level provides a nice visual. It can be seen on Fig. 7.10 that the PICP follows closely to the 1:1 line.

```
cs <- c(99, 97.5, 95, 90, 80, 60, 40, 20, 10, 5)  # confidence level
plot(cs, ((colSums(bMat)/nrow(bMat)) * 100))
abline(a = 0, b = 1, lty = 2, col = "red")
```

From the validation observations the prediction intervals range between 3.2 and 6.4 with a median of about 3.6 pH units when using the Random Forest regression kriging model.

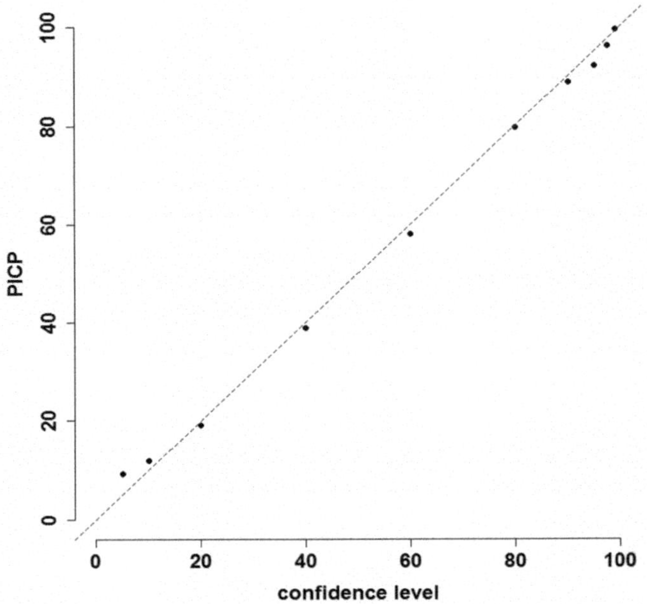

Fig. 7.10 Plot of PICP and confidence level based on validation of Random Forest regression kriging model

```
quantile(uPL.mat[, 4] - lPL.mat[, 4])

##       0%      25%      50%      75%     100%
## 3.193603 3.470226 3.605872 3.964221 6.416089
```

References

Bragato G (2004) Fuzzy continuous classification and spatial interpolation in conventional soil survey for soil mapping of the lower Piave plain. Geoderma 118:1–16

Brown JD, Heuvelink GBM (2005) Assessing uncertainty propagation through physically based models of soil water flow solute transport. In: Encyclopedia of hydrological sciences. John Wiley and Sons, Chichester

Burrough PA, van Gaans PFM, Hootsmans R (1997) Continuous classification in soil survey: spatial correlation, confusion and boundaries. Geoderma 77:115–135

Efron B, Tibshirani R (1993) An introduction to the bootstrap. Chapman and Hall, London

Grimm R, Behrens T (2010) Uncertainty analysis of sample locations within digital soil mapping approaches. Geoderma 155:154–163

Lagacherie P, Cazemier D, vanGaans P, Burrough P (1997) Fuzzy k-means clustering of fields in an elementary catchment and extrapolation to a larger area. Geoderma 77:197–216

Liddicoat C, Maschmedt D, Clifford D, Searle R, Herrmann T, Macdonald L, Baldock J (2015) Predictive mapping of soil organic carbon stocks in south Australia's agricultural zone. Soil Res 53:956–973

Malone BP, McBratney AB, Minasny B (2011) Empirical estimates of uncertainty for mapping continuous depth functions of soil attributes. Geoderma 160:614–626

Malone BP, Minasny B, Odgers NP, McBratney AB (2014) Using model averaging to combine soil property rasters from legacy soil maps and from point data. Geoderma 232–234:34–44

McBratney AB, de Gruijter J (1992) Continuum approach to soil classification by modified fuzzy k-means with extragrades. J Soil Sci 43:159–175

McBratney AB (1992) On variation, uncertainty and informatics in environmental soil management. Aust J Soil Res 30:913–935

McBratney AB, Mendonca Santos ML, Minasny B (2003) On digital soil mapping. Geoderma 117:3–52

Minasny B, McBratney AB (2002a) FuzME version 3.0. Australian Centre for Precision Agriculture, The University of Sydney

Minasny B, McBratney AB (2002b) Uncertainty analysis for pedotransfer functions. Eur J Soil Sci 53:417–429

Odeh I, McBratney AB, Chittleborough D (1992) Soil pattern recognition with fuzzy-c-means: application to classification and soil-landform interrelationships. Soil Sci Soc Am J 56: 506–516

Shrestha DL, Solomatine DP (2006) Machine learning approaches for estimation of prediction interval for the model output. Neural Netw 19:225–235

Solomatine DP, Shrestha DL (2009) A novel method to estimate model uncertainty using machine learning techniques. Water Resour Res 45:Article Number: W00B11

Tranter G, Minasny B, McBratney AB (2010) Estimating pedotransfer function prediction limits using fuzzy k-means with extragrades. Soil Sci Soc Am J 74:1967–1975

Viscarra Rossel RA, Chen C, Grundy MJ, Searle R, Clifford D, Campbell PH (2015) The Australian three-dimensional soil grid: Australia's contribution to the globalsoilmap project. Soil Res 53:845–864

Webster R (2000) Is soil variation random? Geoderma 97:149–163

Chapter 8
Using Digital Soil Mapping to Update, Harmonize and Disaggregate Legacy Soil Maps

Digital soil maps are contrasted from legacy soil maps mainly in terms of the underlying spatial data model. Digital soil maps are based on the pixel data model, while legacy soil maps will typically consist of a tessellation of polygons. The advantage of the pixel model is that the information is spatially explicit. The soil map polygons are delineations of soil mapping units which consist of a defined assemblage of soil classes assumed to exist in more-or-less fixed proportions. There is great value in legacy soil mapping because a huge amount of expertise and resources went into their creation. Digital soil mapping will be the richer by using this existing knowledge-base to derive detailed and high resolution digital soil infrastructures. However the digitization of legacy soil maps is not digital soil mapping. Rather, the incorporation of legacy soil maps into a digital soil mapping workflow involves some method (usually quantitative) of data mining, to appoint spatially explicit soil information—usually a soil class or even a measurable soil attribute—upon a grid the covers the extent of the existing (legacy) mapping. In some ways, this process is akin to downscaling because there is a need to extract soil class or attribute information from aggregated soil mapping units. A better term therefore is soil map disaggregation.

There is an underlying spatial explicitness in digital soil mapping that makes it a powerful medium to portray spatial information. Legacy soil maps also have an underlying spatial model in terms of the delineation of geographical space. However, there is often some subjectivity in the actual arrangement and final shapes of the mapping unit polygons. Yet that is a matter of discussion for another time. For disaggregation studies the biggest impediment to overcome in a quantitative manner is to determine the spatial configuration of the soil classes within each map unit. It is often known which soil classes are in each mapping unit, and sometimes there is information regarding the relative proportions of each too. What is unknown is the spatial explicitness and configuration of said soil classes within the unit. This is the common issue faced in studies seeking the renewal and updating of legacy soil mapping. Some examples of soil map disaggregation studies from

B.P. Malone et al., *Using R for Digital Soil Mapping*,
Progress in Soil Science, DOI 10.1007/978-3-319-44327-0_8

the literature include Thompson et al. (2010) who recovered soil-landscape rules from a soil map report in order to map individual soil classes. This together with a supervised classification approach described by Nauman et al. (2012) represent manually-based approaches to soil map disaggregation. Both of these studies were successfully applied, but because of their manual nature, could also be seen as time-inefficient and susceptible to subjectivity. The flip side to these studies is those using quantitative models. Usually the modeling involves some form of data mining algorithm where knowledge is learned and subsequently optimized based on some model error minimization criteria. Extrapolation of the fitted model is then executed in order to map the disaggregated soil mapping units. Such model-based or data mining procedures for soil map disaggregation include that by Bui and Moran (2001) in Australia, Haring et al. (2012) in Germany and Nauman and Thompson (2014) in the USA. Some fundamental ideas of soil map disaggregation framed in a deeper discussion of scaling of soil information are presented in McBratney (1998).

This chapter seeks to describe a soil map disaggregation method that was first described in Wei et al. (2010) for digital harmonization of adjacent soil surveys in southern Iowa, USA. The concept of harmonization has particular relevance in the USA because it has been long established that the underlying soil mapping concepts across geopolitical boundaries (i.e. counties and states) don't always match. This issue is obviously not a phenomenon exclusive to the USA but is a common worldwide issue. This mismatch may include the line drawings and named map units. Of course, soils in the field do not change at these political boundaries. These soil-to-soil mismatches are the result of the past structuring of the soil survey program. For example, soil surveys in the US were conducted on a soil survey area basis. Most times the soil surveys areas were based on county boundaries. Often adjacent counties were mapped years apart. Different personnel, different philosophies of soil survey science, new concepts of mapping and the availability of various technologies all have played a part in why these differences occur. These differences maybe even more exaggerated at state lines as each state was administratively and technically responsible for the soil survey program within a given state. The algorithm developed by Wei et al. (2010) addressed this issue, where soil mapping units were disaggregated into soil series. Instead of mapping the prediction of a single soil series, a probability distribution of all potential soil series was estimated. The outcome of this was the dual disaggregation and harmonization of existing legacy soil information into raster-based digital soil mapping product/s.

Odgers et al. (2014) using legacy soil mapping from an area in Queensland, Australia refined the algorithm to which they called DSMART or, Disaggregation and Harmonization of Soil Map Units Through Resampled Classification Trees. Besides the work of Odgers et al. (2014), The DSMART algorithm has been used is other projects throughout the world, with Chaney et al. (2014) using it to disaggregate the entire gridded USA Soil Survey Geographic (gSSURGO) database. The resulting POLARIS data set (Probabilistic Remapping of SSURGO) provides the 50 most probable soil series predictions at each 30-meter grid cell over the contiguous USA. DSMART has also been a critical component for the development of the Soil and Landscape Grid of Australia (SLGA) data set (Grundy et al. 2015). The SLGA is the first continental version of the GlobalSoilMap.net concept and the first nationally consistent, fine spatial resolution set of continuous soil attributes with Australia-wide coverage. The DSMART algorithm has been pivotal, together with the associated PROPR algorithm (Digital Soil Property Mapping Using Soil Class Probability Rasters; Odgers et al. (2015)) in deriving high resolution digital soil maps where point-based DSM approaches cannot be undertaken, particularly where soil point data is sparse. In this chapter, the fundamental features of DSMART are described, followed its demonstration upon a small data set.

8.1 DSMART: An Overview

Odgers et al. (2014) provide a detailed explanation of the DSMART algorithm. The aim of DSMART is to predict the spatial distribution of soil classes by disaggregating the soil map units of a soil polygon map. Here soil map units soil map are entities consisting of a defined set of soil classes which occur together in a certain spatial pattern and in an assumed set of proportions. The DSMART method of representing the disaggregated soil class distribution is as a set of numerical raster surfaces, with one raster per soil class. The data representation for each soil class is given as the probability of occurrence. In order to generate the probability surfaces, a re-sampling approach is used to generate *n* realizations of the potential soil class distribution within each map unit. Then at each grid cell, the probability of occurrence of each soil class is estimated by the proportion of times the grid cell is predicted as each soil class across the set of realizations. The procedure of the DSMART algorithm can be summarized in 6 main steps:

1. Draw *n* random samples from each soil map polygon.
2. Assign soil class to each sampling point.
    - Weighted random allocation from soil classes in relevant map unit
    - Relative proportions of soil classes within map units are used as the weights
3. Use sampling points and intersected covariate values to build a decision tree to prediction spatial distribution of soil classes.
4. Apply decision tree across mapping extent using covariate layers
5. Steps 1–4 repeated *i* times to produce *i* realizations of soil class distribution.
6. Using *i* realizations generate probability surfaces for each soil class.

The model type that Odgers et al. (2014) used was the C4.5 decision tree algorithm which was introduced by Quinlan (1993). The type of data mining algorithm implemented in DSMART is not prescriptive; as long as it is robust and importantly, computationally efficient. For example Chaney et al. (2014) used Random Forest models (Breiman 2001) in their implementation of DSMART.

8.2 Implementation of DSMART

The DSMART algorithm has previously been written in the C++ and Python computing languages. It is also available in an R package, which was developed at the Soil Security Laboratory. Regardless of computing language preference, DSMART requires three chief sources of data:

1. The soil map unit polygons that will be disaggregated.
2. Information about the soil class composition of the soil map unit polygons
3. Geo-referenced raster covariates representing the *scorpan* factors of which have complete and continuous coverage of the mapping extent. There is no restriction in terms of the data type i.e. continuous, categorical, ordinal etc.

8.2.1 DSMART with R

The DSMART R package contains two working functions: `dsmart` and `dsmartR`. More will be discussed about these shortly. The other items in the package are various inputs required to run the function. In essence these data provide some indication of the structure and nature of the information that is required to run the DSMART algorithm so that it can be easily adapted to other project. First is the soil map to be disaggregated. This is saved to the `dsT_polygons` object. In this example the small polygon map is a clipped area of the much larger soil map that Odgers et al. (2014) disaggregated, which was the 1:250,000 soil map of the Dalrymple Shire in Queensland Australia by Rogers et al. (1999).In this example data set, there are 11 soil mapping units.the polygon object is of class `SpatialPolygonsDataFrame`, which is what would be created if you were to read in a shapefile of the polygons into R (Fig. 8.1).

```
library(dsmart) install_bitbucket("brendo1001/dsmart/rPackage/
   dsmart/pkg")
library(devtools)
library(sp)
library(raster)

# Polygons
data(dsT_polygons)
class(dsT_polygons)

## [1] "SpatialPolygonsDataFrame"
## attr(,"package")
## [1] "sp"

summary(dsT_polygons$MAP_CODE)

## BUGA1t CGCO3t   DO3n   FL3d   HG2g   MI6t   MM5g   MS4g   PA1f   RA3t
##      1      1      1      1      1      1      1      1      1      1
```

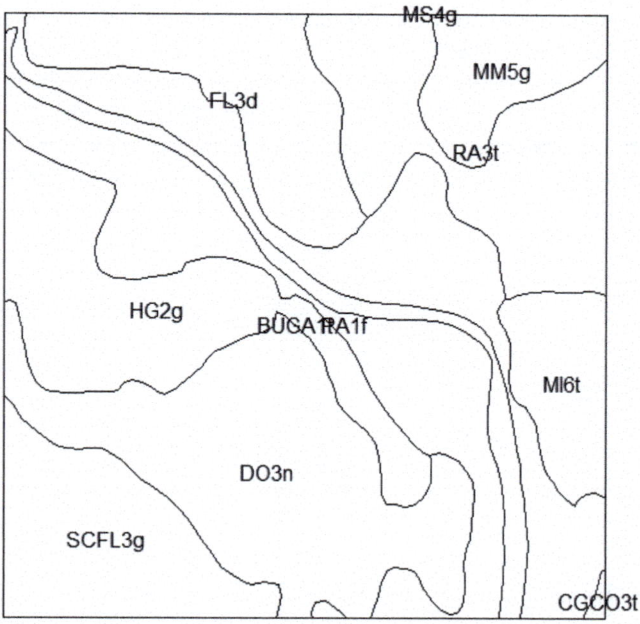

Fig. 8.1 Subset of the polygon soil map from the Dalrymple Shire, Queensland Australia which was disaggregated by Odgers et al. (2014)

```
## SCFL3g
##      1
```

```
plot(dsT_polygons)
invisible(text(getSpPPolygonsLabptSlots(dsT_polygons),
labels = as.character(dsT_polygons$MAP_CODE),cex = 1))
```

The next inputs are the soil map unit compositions which is saved to the `dsT_composition` object. This is a data frame that simply indicates in respective columns the map unit name, and corresponding numerical identifier label. Then there is the soil classes that are contained in the respective mapping unit, followed by the relative proportion that each soil class contributes to the map unit. The relative proportions will and probably should sum to 100.

```
# Map unit compositions
data(dsT_composition)
head(dsT_composition)
```

```
##   poly mapunit soil_class proportion
## 1  304    MM5g         RA         70
## 2  304    MM5g         EW         20
## 3  304    MM5g         PI         10
## 4  440  CGCO3t         CG         50
## 5  440  CGCO3t         CO         20
## 6  440  CGCO3t         DA         10
```

The last required inputs are the environmental covariates. This use used to inform the model fitting for each DSMART iteration, and ultimately be used for the spatial mapping. There are actually 20 different covariate rasters of which have been derived from a digital elevation model and gamma radiometric data. These rasters are organized into a `RasterStack` and are of 30 m grid resolution. This class of data is the necessary format of the covariate data for input into DSMART.

```
# covariates
data(dsT_covariates)
class(dsT_covariates)

## [1] "RasterStack"
## attr(,"package")
## [1] "raster"

nlayers(dsT_covariates)

## [1] 20

res(dsT_covariates)

## [1] 30 30
```

Now it is time to run the DSMART algorithm. The actual `R` implementation is spread across two companion functions already mentioned: `dsmart` and `dsmartR`. The `dsmart` function is the workhorse of the two because it performs the sample/resampling, model fitting and iteration parts of DSMART. The `dsmartR` function works on the outputs of `dsmart` to estimate the probabilities of classes, and derives some other useful outputs such as the most probable soil class and or n-most probable soil classes. Using the `dsmart` function, we provide it will the inputs described above. Additional inputs include the parameters `n`, which is a numeric value for the number of samples to take from each soil mapping polygon; `reals` is the number of model realizations to fit; and `cpus` is the number of compute nodes to use for the analysis. The default is to run the algorithm in sequential mode, however it does have the capability to be scaled up substantially in parallel mode, which helps to eliminate some computation time. In the example below we set `n` to 15, `reals` to 10, and `cpus` to 4. An unusual feature of this function is that none of the output is saved to the `R` memory, but instead, directly to file, or specifically the current working directory into a folder called `dsmartOuts`. After the `dsmart` function has terminated, you will find in the `dsmartOuts` folder a few other folders which contain rasters of the soil class prediction from each iteration. You will also encounter another folder contain text file outputs from each iteration of the C5 model structure plus information on the quality of the fit. Each model is saved to the `dsmartModels.RData` object that is also found in the outputs folder.

```
# run dsmart
library(parallel)
test.dsmart <- dsmart(covariates = dsT_covariates,
polygons = dsT_polygons,composition = dsT_composition, n = 15,
reals = 10, cpus = 4)
```

Of particular interest is to derive the soil class probabilities, and even the most probable soil class at each pixel, and even an estimate of the uncertainty, which is given in terms of the confusion index that was used earlier during the fuzzy classification of data for derivation of digital soil map uncertainties. The confusion index essentially measure how similar to classification is between (in most cases) most probable and second-most probable soil class predictions at a pixel. To derive the probability rasters we need the rasters that were generated from the `dsmart` function. This can be done via the use of the `list.files` function and the `raster` package to read in the rasters and stack them into a `rasterStack`. Rather than doing this we can use pre-prepared outputs namely in the form of the `dsmartOutMaps` (raster outputs from `dsmart`) and `dsT_lookup` (lookup table of soil class names and numeric counterparts) objects. As with `dsmart` this function can be run in parallel mode via control of the `cpus` variable. A logical entry is required for the `sepP` variable to determine if probability rasters are required t be created for each soil class. These particular outputs are important for the follow on procedure of soil attribute mapping, of which is the focus of the study by Odgers et al. (2015) and integral to the associated PROPR algorithm. In many cases the user may just be interested in deriving the most probable soil class, or sometimes the n-most probable soil class maps.

```
# run dsmartR run function getting most probable and creating
# probability
# rasters using 4 compute cores.
data(dsmartOutMaps)
data(dsT_lookup)
test.dsmartR <- dsmartR(rLocs = dsmartOutMaps, nprob = 1,
sepP = TRUE, lookup = dsT_lookup, cpus = 4)
```

A few folders are automatically generated to the working directory which contain various outputs from the `dsmartR` function. These are: `nProbable`, `probabilities`, and `counts`. The folder `nProbable` contains the n-most probable soil class maps. If requested, the probabilities of the n-most probable, together with an estimated confusion index (Burrough et al. 1997) is returned. The folder: `probabilities`, contains the probability rasters for each candidate soil class if requested. The folder: `counts`, contains a `RasterBrick` of the count of predicted candidate soil classes from the output realizations of dsmart. A `list` of the n-most probable rasters, and if requested; probability rasters, n-most probable probability rasters and confusion index raster are saved to the working memory. So as a final step, lets produce a map of the most probable soil class (Fig. 8.2) , and the associated map of the confusion index (Fig. 8.3).

```
# plot most probable soil class
library(rasterVis)
ml.map <- test.dsmartR[[2]][[1]]
ml.map
ml.map <- as.factor(ml.map)
rat <- levels(ml.map)[[1]]
```

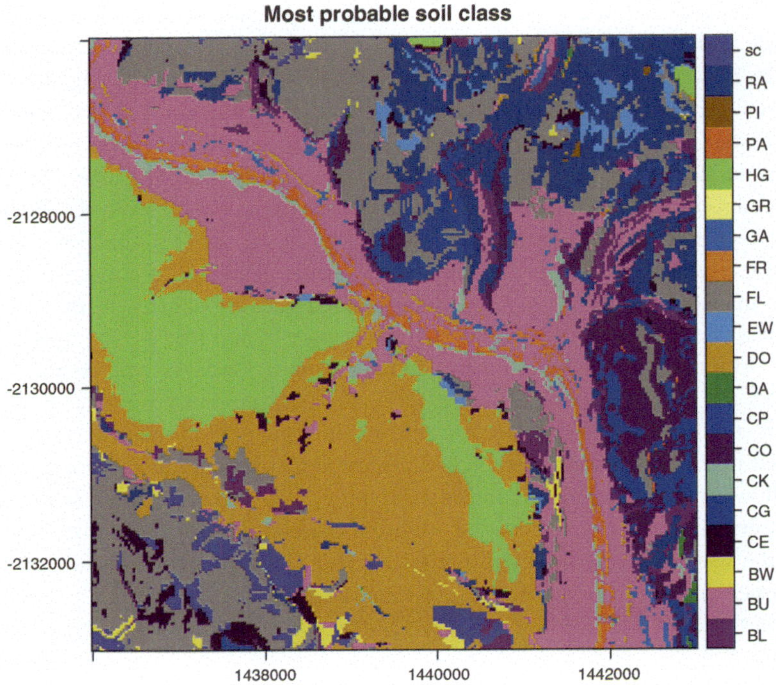

Fig. 8.2 Map of the most probable soil class from DSMART

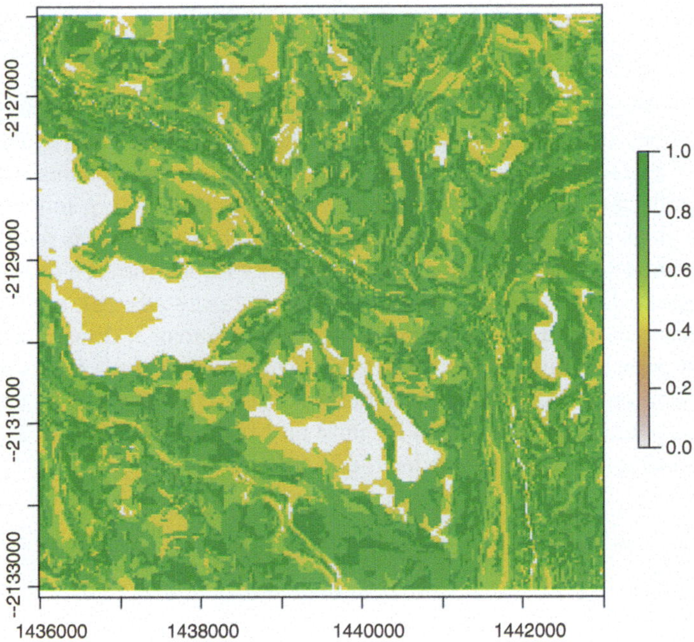

Fig. 8.3 Confusion index of soil class predictions from DSMART

```
rat[["class"]] <- c("BL", "BU", "BW", "CE", "CG", "CK", "CO",
"CP", "DA", "DO", "EW", "FL", "FR", "GA", "GR", "HG", "PA", "PI",
"RA", "SC")
levels(ml.map) <- rat

# Randomly selected HEX colors
area_colors <- c("#9a10a8", "#cf68a0", "#e7cc15", "#4f043a",
"#1129a3", "#a2d0a7", "#7b0687", "#3e28d1", "#2c8c04", "#d39014",
"#66a5ed", "#978279", "#db6f1f", "#4070fc", "#fde864", "#acdb0a",
"#d95a28", "#94561f", "#162972", "#8342e1")
levelplot(ml.map, col.regions = area_colors, xlab = "", ylab = "",
main = "Most probable soil class")

# Confusion Index
CI.map <- test.dsmartR[[4]]
plot(CI.map)
```

DSMART can be quite a powerful algorithm for disaggregating legacy soil mapping as demonstrated by Chaney et al. (2014) in the USA. While the computational effort to generate disaggregated predictions could be a burden, the DSMART algorithm is relatively easy to parallelize, which was also demonstrated in Chaney et al. (2014), and with the implementation of the current R version of this algorithm. One other restrictive feature of the DSMART algorithm is the need for specific inputs, particularly in regards to the soil class compositions and their relative proportions within mapping units. Sometimes this information is not easily available, and needs to be approximated by some means.

Besides to advantage of generating detailed maps of soil classes, which will have their own use for some applications, the DSMART algorithm provides a pathway to first update and harmonize legacy soil maps, and then to realize soil property information from existing polygon soil maps, such as through the use and coupling of soil class probability rasters and modal soil class profiles as demonstrated in Odgers et al. (2015). It is this detailed soil attribute information that is required in land system modeling frameworks, and for ongoing assessment and monitoring of the soil resource.

References

Breiman L (2001) Random forests. Mach Learn 41:5–32

Bui E, Moran C (2001) Disaggregation of polygons of surficial geology and soil maps using spatial modelling and legacy data. Geoderma 103:79–94

Burrough PA, van Gaans PFM, Hootsmans R (1997) Continuous classification in soil survey: spatial correlation, confusion and boundaries. Geoderma 77:115–135

Chaney N, Hempel JW, Odgers NP, McBratney AB, Wood EF (2014) Spatial disaggregation and harmonization of gSSURGO. In: ASA, CSSA and SSSA international annual meeting, Long Beach. ASA, CSSA and SSSA

Grundy MJ, Viscarra Rossel R, Searle RD, Wilson PL, Chen C, Gregory LJ (2015) Soil and landscape grid of Australia. Soil Res. http://dx.doi.org/10.1071/SR15191

Haring T, Dietz E, Osenstetter S, Koschitzki T, Schroder B (2012) Spatial disaggregation of complex soil map units: a decision-tree based approach in Bavarian forest soils. Geoderma 185–186:37–47

McBratney A (1998) Some considerations on methods for spatially aggregating and disaggregating soil information. In: Finke P, Bouma J, Hoosbeek M (eds) Soil and water quality at different scales. Developments in plant and soil sciences, vol 80. Springer, Dordrecht, pp 51–62

Nauman TW, Thompson JA (2014) Semi-automated disaggregation of conventional soil maps using knowledge driven data mining and classification trees. Geoderma 213:385–399

Nauman TW, Thompson JA, Odgers NP, Libohova Z (2012) Fuzzy disaggregation of conventional soil maps using database knowledge extraction to produce soil property maps. In: Digital soil assessments and beyond: Proceedings of the fifth global workshop on digital soil mapping. CRC Press, London, pp 203–207

Odgers NP, McBratney AB, Minasny B (2015) Digital soil property mapping and uncertainty estimation using soil class probability rasters. Geoderma 237–238:190–198

Odgers NP, Sun W, McBratney AB, Minasny B, Clifford D (2014) Disaggregating and harmonising soil map units through resampled classification trees. Geoderma 214–215:91–100

Quinlan JR (1993) C4.5: programs for machine learning. Morgan Kaufmann, San Mateo

Rogers L, Cannon M, Barry E (1999) Land resources of the Dalrymple shire, 1. Land resources bulletin DNRQ980090. Department of Natural Resources, Brisbane, Queensland

Thompson JA, Prescott T, Moore AC, Bell J, Kautz DR, Hempel JW, Waltman SW, Perry C (2010) Regional approach to soil property mapping using legacy data and spatial disaggregation techniques. In: 19th world congress of soil science. IUSS, Brisbane

Wei S, McBratney A, Hempel J, Minasny B, Malone B, D'Avello T, Burras L, Thompson J (2010) Digital harmonisation of adjacent analogue soil survey areas – 4 Iowa counties. In: 19th world congress of soil science, IUSS, Brisbane

Chapter 9
Combining Continuous and Categorical Modeling: Digital Soil Mapping of Soil Horizons and Their Depths

The motivation for this chapter is to gain some insights into a digital soil mapping approach that uses a combination of both continuous and categorical attribute modeling. Subsequently, we will build on the efforts of the material in the chapters that dealt with each of these type of modeling approaches separately. There are some situations where, a combinatorial approach might be suitable in a digital soil mapping work flow.

An example of such a workflow is in Malone et al. (2015) in regards to the prediction of soil depth. The context behind that approach was that often lithic contact was not achieved during the soil survey activities, effectively indicating soil depth was greater than the soil probing depth (which was 1.5 m). Where lithic contact was found, the resulting depth was recorded. The difficulty in using this data in the raw form was that there were many sites where soil depth was greater than 1.5 m together with actual recorded soil depth measurements. The nature of this type of data is likened to a zero-inflated distribution, where many zero observations are recorded among actual measurements (Sileshi 2008). In Malone et al. (2015) the zero observations were attributed to soil depth being greater than 1.5 m. They therefore performed the modeling in two stages. First modeling involved a categorical or binomial model of soil depth being greater than 1.5 m or not. This was followed by continuous attribute modeling of soil depth using the observations where lithic contact was recorded. While the approach was a reasonable solution, it may be the case that the frequency of recorded measurements is low, meaning that the spatial modeling of the continuous attribute is made under considerable uncertainty, as was the case in Malone et al. (2015) with soil depth and other environmental variables spatially modeled in that study; for example, the frequency of winter frosts.

Another interesting example of a combinatorial DSM work flow was described in Gastaldi et al. (2012) for the mapping of occurrence and thickness of soil profiles. There they used a multinomial logistic model to predict the presence or absence of the given soil horizon class, followed by continuous attribute modeling of the hori-

B.P. Malone et al., *Using R for Digital Soil Mapping*,
Progress in Soil Science, DOI 10.1007/978-3-319-44327-0_9

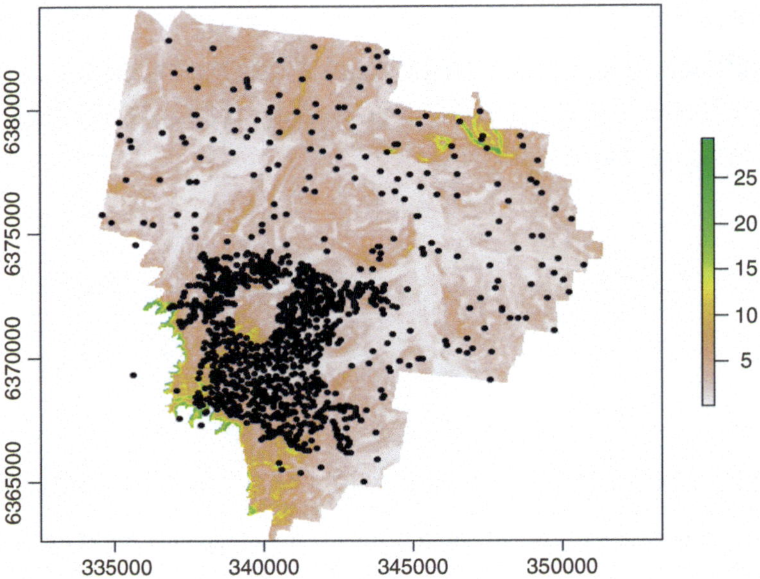

Fig. 9.1 Hunter Valley soil profiles locations overlaying digital elevation model

zon depths. For the purposes a demonstrating the work flow of this combinatorial or two-stage DSM, we will re-visit the approach that is described by Gastaldi et al. (2012) and work through the various steps needed to perform it within R.

The data we will use comes from 1342 soil profile and core descriptions from the Lower Hunter Valley, NSW Australia. These data have been collected on an annual basis since 2001 to present. These data are distributed across the 220 km^2 area as shown in Fig. 9.1. The intention is to use these data first to predict the occurrence of given soil horizon classes (following the nomenclature of the Australian Soil Classification (Isbell 2002)). Specifically we want to prediction the spatial distribution of the occurrence of A1, A2, AP, B1, B21, B22, B23, B24, BC, and C horizons, and then where those horizons occur, predict their depth.

First lets perform some data discovery both in terms of the soil profile data and spatial covariates to be used as predictor variables and to inform the spatial mapping. You will notice the soil profile data `dat` is arranged in a flat file where each row is a soil profile. There are many columns of information which include profile identifier and spatial coordinates. Then there are 11 further columns that are binary indicators of whether a horizon class is present or not (indicated as 1 and 0 respectively). The following 11 columns after the binary columns indicate the horizon depth for the given horizon class.

```
library(sp)
library(raster)
# data
str(dat)
```

```
## 'data.frame': 1342 obs. of  25 variables:
##  $ FID : Factor w/ 1342 levels "1","10","100",..: 1 2 3 ...
##  $ e   : num  338014 338183 341609 341352 339736 ...
##  $ n   : num  6370646 6370550 6370437 6370447 6370439 ...
##  $ A1  : int  1 0 1 1 1 1 1 0 1 1 ...
##  $ A2  : int  1 0 0 0 0 1 1 0 0 0 ...
##  $ AP  : int  0 1 0 0 0 0 0 1 0 0 ...
##  $ B1  : int  0 0 1 1 1 0 0 0 1 0 ...
##  $ B21 : int  1 1 1 1 1 1 1 1 1 1 ...
##  $ B22 : int  1 0 1 0 1 1 1 1 1 1 ...
##  $ B23 : int  0 0 0 0 0 0 0 0 0 0 ...
##  $ B24 : int  0 0 0 0 0 0 0 0 0 0 ...
##  $ B3  : int  0 0 0 0 0 0 0 0 0 0 ...
##  $ BC  : int  0 0 0 0 0 0 0 0 1 1 ...
##  $ C   : int  0 0 0 0 0 0 0 0 0 0 ...
##  $ A1d : num  21 NA 17 45 20 25 13 NA 10 44 ...
##  $ A2d : num  27 NA NA NA NA 15 13 NA NA NA ...
##  $ APd : num  NA 40 NA NA NA NA NA 35 NA NA ...
##  $ B1d : num  NA NA 25 25 30 NA NA NA 30 NA ...
##  $ B21d: num  26 60 26 30 30 25 58 40 20 38 ...
##  $ B22d: num  26 NA 32 NA 20 20 16 25 25 18 ...
##  $ B23d: int  NA NA NA NA NA NA NA NA NA NA ...
##  $ B24d: int  NA NA NA NA NA NA NA NA NA NA ...
##  $ B3d : int  NA NA NA NA NA NA NA NA NA NA ...
##  $ BCd : int  NA NA NA NA NA NA NA NA 15 NA ...
##  $ Cd  : int  NA NA NA NA NA NA NA NA NA NA ...
```

```
# convert data to spatial object
coordinates(dat) <- ~e + n
```

At our disposal are a number of spatial covariates that have either been derived from a digital elevation model, airborne gamma radiometric survey and Landsat satellite spectral wavelengths. These are all registered to the common spatial resolution of 25m and have been organized together into a `rasterStack`.

```
# covariates
names(s1)
```

```
##  [1] "totalCount"                          "thppm"
##  [3] "Terrain_Ruggedness_Index"            "slope"
##  [5] "SAGA_wetness_index"                  "r57"
##  [7] "r37"                                 "r32"
##  [9] "PC2"                                 "PC1"
## [11] "ndvi"                                "MRVBF"
## [13] "MRRTF"                               "Mid_Slope_Positon"
## [15] "light_insolation"                    "kperc"
## [17] "Filled_DEM"                          "drainage_2011"
## [19] "Altitude_Above_Channel_Network"
```

```
# resolution
res(s1)
```

```
## [1] 25 25

# raster properties
dim(s1)

## [1] 860 675  19
```

For a quick check, lets overlay the soil profile points onto the DEM. You will notice on Fig. 9.1 the area of concentrated soil survey (which represents locations of annual survey) within the extent of a regional scale soil survey across the whole study area.

```
plot(raster(files[17]))
points(dat, pch = 20)
```

The last preparatory step we need to take is the covariate intersection of the soil profile data, and remove any sites that are outside the mapping extent.

```
# Covariate extract
ext <- extract(s1, dat, df = T, method = "simple")
w.dat <- cbind(as.data.frame(dat), ext)

# remove sites with missing covariates
x.dat <- w.dat[complete.cases(w.dat[, 27:45]), ]
```

9.1 Two-Stage Model Fitting and Validation

A demonstration will be given of the two-stage modeling work flow for the A1 horizon, but given some indication of the results for the other horizons and their depths further on. First we want to subset 75 % of the data for calibrating models, and keeping the rest aside for validation purposes.

```
# A1 Horizon
x.dat$A1 <- as.factor(x.dat$A1)
# random subset
set.seed(123)
training <- sample(nrow(x.dat), 0.75 * nrow(x.dat))
# calibration dataset
dat.C <- x.dat[training, ]
# validation dataset
dat.V <- x.dat[-training, ]
```

We first want to model the presence/absence in this case of the A1 horizon. We will use a multinomial model, followed up with a stepwise regression procedure in order to remove non-significant predictor variables.

```
library(nnet)
library(MASS)


# A1 presence or absence model
mn1 <- multinom(formula = A1 ~ totalCount + thppm +
Terrain_Ruggedness_Index + slope + SAGA_wetness_index + r57 +
r37 + r32 + PC2 + PC1 + ndvi + MRVBF + MRRTF + Mid_Slope_Positon +
light_insolation + kperc + Filled_DEM + drainage_2011 +
Altitude_Above_Channel_Network, data = dat.C)

# stepwise variable selection
mn2 <- stepAIC(mn1, direction = "both", trace = FALSE)


summary(mn2)

## Call:
## multinom(formula = A1 ~ SAGA_wetness_index + r57 + r37 + r32 +
##     Filled_DEM + Altitude_Above_Channel_Network, data = dat.C)
##
## Coefficients:
##                                    Values  Std. Err.
## (Intercept)                    -6.77776938 2.25295276
## SAGA_wetness_index              0.16402675 0.06413479
## r57                             5.19885165 0.62748971
## r37                            -3.18245854 1.50679841
## r32                            -4.22725150 1.15143384
## Filled_DEM                      0.02983971 0.00503181
## Altitude_Above_Channel_Network -0.03387555 0.01051877
##
## Residual Deviance: 629.0489
## AIC: 643.0489
```

We use the goofcat function from ithir to assess the model quality both in terms of the calibration and validation data.

```
# calibration
mod.pred <- predict(mn2, newdata = dat.C, type = "class")
goofcat(observed = dat.C$A1, predicted = mod.pred)

## $confusion_matrix
##     0   1
## 0  28  20
## 1 109 841
##
## $overall_accuracy
## [1] 88
##
## $producers_accuracy
##  0  1
## 21 98
```

```
##
## $users_accuracy
##  0  1
## 59 89
##
## $kappa
## [1] 0.2492215
```

```
# validation
val.pred <- predict(mn2, newdata = dat.V, type = "class")
goofcat(observed = dat.V$A1, predicted = val.pred)
```

```
## $confusion_matrix
##     0   1
## 0   7   6
## 1  38 282
##
## $overall_accuracy
## [1] 87
##
## $producers_accuracy
##  0  1
## 16 98
##
## $users_accuracy
##  0  1
## 54 89
##
## $kappa
## [1] 0.1924603
```

It is clear that the `mn2` model is not too effective for predicting sites where the A1 horizon is absent.

What we want to do now is to model the A1 horizon depth. We will be using an alternative model to those that have been examined in this book so far. The is a quantile regression forest, which is a generalized implementation of the random forest model from Breiman (2001). The algorithm is available via the `quantregForest` package, and further details about the model can be found at Meinshausen (2006). The `caret` package also interfaces with this model too. Fundamentally, random forests are integral to the quantile regression algorithm. However, the useful feature and advancement from normal random forests is the ability to infer the full conditional distribution of a response variable. This facility is useful for building non-parametric prediction intervals for a any given level of confidence information and also the ability to detect outliers in the data easily. Quantile regression used via the `quanregForest` algorithm is implemented in the chapter simply to demonstrate the wide availability of prediction models and machine learning methods that can be used in digital soil mapping.

Getting the model initiated we first need to perform some preparatory tasks. Namely the removal of missing data from the available data set.

```
# Remove missing values calibration
mod.dat <- dat.C[!is.na(dat.C$A1d), ]  #calibration

# validation
val.dat <- dat.V[!is.na(dat.V$A1d), ]
```

It is useful to check the inputs required for the quantile regression forests (using the help file); however its parameterization is largely similar to other models that have been used already in this book, particularly those for the random forest models.

```
# Fit quantile regression forest
library(quantregForest)
qrf <- quantregForest(x = mod.dat[, 27:45], y = mod.dat$A1d,
importance = TRUE)
```

Before we use the `goof` function to assess the model quality, a very helpful graphical output from the model is the plot of the out-of-bag samples with respect to whether the measured values are inside or outside their prediction interval (Fig. 9.2. Recall that the out-of-bag samples are those that are not included in the regression forest model iterations. Further note that it is also possible to define prediction intervals to your own wishing. The default output is for a 90 % prediction interval.

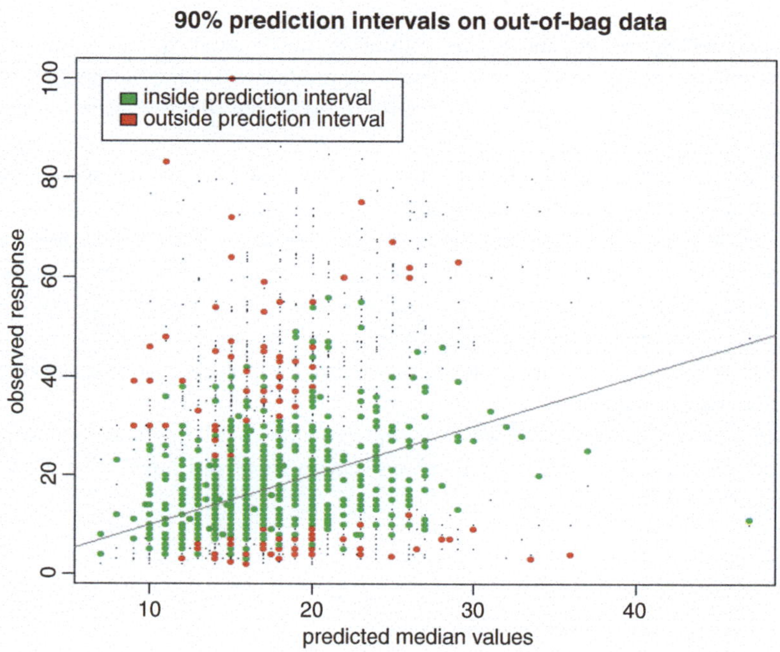

Fig. 9.2 90 % prediction intervals on out-of-bag data for predicting depth of A1 horizon

```
plot(qrf)
```

Naturally, the best test of the model is to use an external data set. In addition to our normal validation procedure we can also derive the PICP for the validation data too.

```
## Calibration
quant.cal <- predict(qrf, newdata = mod.dat[, 27:45], all = T)
goof(observed = mod.dat$A1d, predicted = quant.cal[, 2])

##          R2 concordance      MSE     RMSE      bias
## 1 0.6541884   0.4829592 82.17118 9.064832 -2.354085

# Validation
quant.val <- predict(qrf, newdata = val.dat[, 27:45], all = T)
goof(observed = val.dat$A1d, predicted = quant.val[, 2])

##           R2 concordance      MSE     RMSE      bias
## 1 0.01192341  0.05000794 112.7926 10.62039 -1.542385

# PICP
sum(quant.val[, 1] <= val.dat$A1d & quant.val[, 2] >=
val.dat$A1d)/nrow(val.dat)

## [1] 0.4479167
```

Based on the outputs above, the calibration model seems a reasonable outcome for the model, but is proven to be largely un-predictive for the validation data set. We should also be expecting a PICP close to 90 %, but this is clearly not the case above.

What has been covered above for the two-stage modeling is repeated for all the other soil horizons, with the main results displayed in Table 9.1. These statistics are reported based on the validation data. It clear that there is a considerable amount of uncertainty overall in the various soil horizon models. For some horizons the results are a little encouraging; for example the model to predict the presence of a BC horizon is quite good. It is clear however that distinguishing between different B2 horizons is challenging. However predicting the presence or absence of a B22 horizons seems acceptable.

Another way to assess the quality of the two-stage modeling is to assess first the number of soil profile that have matching sequences of soil horizon types. We can do this using:

```
# Validation data horizon observations (1st 3 rows)
dat.V[1:3, c(1, 4:14)]

##     FID A1 A2 AP B1 B21 B22 B23 B24 B3 BC C
## 1   101  1  0  0  1   1   0   0   0  0  0 0
## 2  1022  0  0  1  0   1   1   0   0  0  0 0
## 3  1026  1  0  0  0   1   1   0   0  0  0 0
```

Table 9.1 Selected model validation diagnostics returned for each horizon class and associated depth model

	Presence/Absence of horizon			Depth of horizon		
Horizon	Overall accuracy	User's accuracy	Kappa statistic	Concordance	RMSE	PICP
A1	87 %	Pres = 89 % Abs = 54 %	0.19	0.05	10	46 %
A2	87 %	Pres = 100 % Abs = 87 %	0.04	0.10	12	42 %
AP	86 %	Pres = 50 Abs = 88 %	0.15	0.00	12	53 %
B1	91 %	Pres = 0 Abs = 91 %	0	0.16	12	45 %
B21	97 %	Pres = 97 Abs = 0 %	0	0.05	17	41 %
B22	73 %	Pres = 73 Abs = 34 %	0	0.10	14	41 %
B23	78 %	Pres = 0 Abs = 78 %	0	0.04	12	45 %
B24	97 %	Pres = 0 Abs = 97 %	0	0.00	22	46 %
BC	74 %	Pres = 68 Abs = 75 %	0.20	0.06	18	29 %
C	95 %	Pres = 0 Abs = 95 %	0	0	NA	68 %

```
# Associated model predictions (1st 3 rows)
vv.dat[1:3, 1:12]

##    dat.V.FID a1 a2 ap b1 b21 b22 b23 b24 b3 bc c
## 1        101  1  0  0  0   1   1   0   0  0  0 0
## 2       1022  1  0  0  0   1   1   0   0  0  0 0
## 3       1026  1  0  0  0   1   1   0   0  0  0 0

# matched soil profiles
sum(dat.V$A1 == vv.dat$a1 & dat.V$A2 == vv.dat$a2 & dat.V$AP == vv.dat$ap &
dat.V$B1 == vv.dat$b1 & dat.V$B21 == vv.dat$b21 & dat.V$B22 == vv.dat$b22 &
dat.V$B23 == vv.dat$b23 & dat.V$B24 == vv.dat$b24 & dat.V$B3 == vv.dat$b3 &
dat.V$BC == vv.dat$bc & dat.V$C == vv.dat$c)/nrow(dat.V)

## [1] 0.2222222
```

The result above indicates that just over 20 % of validation soil profiles have matched sequences of horizons. We can examine visually a few of these matched profiles to examine whether there is much coherence in terms of observed and associated predicted horizon depths. We will select out two soil profiles: One with an AP horizon, and the other with an A1 horizon. We can demonstrate this using the aqp package, which is a dedicated R package for handling soil profile data collections.

```
# Subset of matching data (observations)
match.dat <- dat.V[which(dat.V$A1 == vv.dat$a1 & dat.V$A2 == vv.dat$a2 &
dat.V$AP == vv.dat$ap & dat.V$B1 == vv.dat$b1 & dat.V$B21 == vv.dat$b21 &
dat.V$B22 == vv.dat$b22 & dat.V$B23 == vv.dat$b23 & dat.V$B24 == vv.dat$b24 &
dat.V$B3 == vv.dat$b3 & dat.V$BC == vv.dat$bc & dat.V$C == vv.dat$c), ]

# Subset of matching data (predictions)
match.dat.P <- vv.dat[which(dat.V$A1 == vv.dat$a1 & dat.V$A2 == vv.dat$a2 &
dat.V$AP == vv.dat$ap & dat.V$B1 == vv.dat$b1 & dat.V$B21 == vv.dat$b21 &
dat.V$B22 == vv.dat$b22 & dat.V$B23 == vv.dat$b23 & dat.V$B24 == vv.dat$b24 &
dat.V$B3 == vv.dat$b3 & dat.V$BC == vv.dat$bc & dat.V$C == vv.dat$c), ]
```

Now we just select any row where we know there is and AP horizon

```
match.dat[49, ]  #observation

##     FID      e       n A1 A2 AP B1 B21 B22 B23 B24 B3 BC C A1d A2d APd B1d
## 195 642 338096 6372259  0  0  1  0   1   1   0   0  0  1 0  NA  NA  10  NA
##     B21d B22d B23d B24d B3d BCd Cd  ID totalCount    thppm
## 195   30   15   NA   NA  NA  45 NA 735   446.7597 7.192239
##     Terrain_Ruggedness_Index    slope SAGA_wetness_index      r57      r37
## 195                 0.846727 0.697118           13.34301 1.955882 0.794118
##          r32      PC2       PC1      ndvi    MRVBF    MRRTF
## 195 1.542857 -1.89351 -2.239939 -0.076923 0.111123 3.746326
##     Mid_Slope_Positon light_insolation     kperc Filled_DEM drainage_2011
## 195          0.130692         1716.388 0.5863795   142.8293      3.909594
##     Altitude_Above_Channel_Network
## 195                       25.52147

match.dat.P[49, ]  #prediction

##     dat.V.FID a1 a2 ap b1 b21 b22 b23 b24 b3 bc c a1d a2d      apd b1.1
## 195       642  0  0  1  0   1   1   0   0  0  1 0  18  16 21.92308   18
##     b21d b22d b23d     b24d b3d bcd cd
## 195   31   27   20 15.41176  NA  32 NA
```

We can see in these two profiles, the sequence of horizons is AP, B21, B22, BC. Now we just need to upgrade the data to a soil profile collection. Using the horizon classes together with the associated depths, we want to plot both soil profiles for comparison. First we need to create a data frame of the relevant data then upgrade to a `soilProfileCollection`, then finally plot. The script below demonstrates this for the soil profile with the AP horizon. The same can be done with the associated soil profile with the A1 horizon. The plot of the is shown on Fig. 9.3.

```
# Horizon classes
H1 <- c("AP", "B21", "B22", "BC")

# Extract horizon depths then combine to create soil profiles
p1 <- c(22, 31, 27, 32)
p2 <- c(10, 30, 15, 45)
```

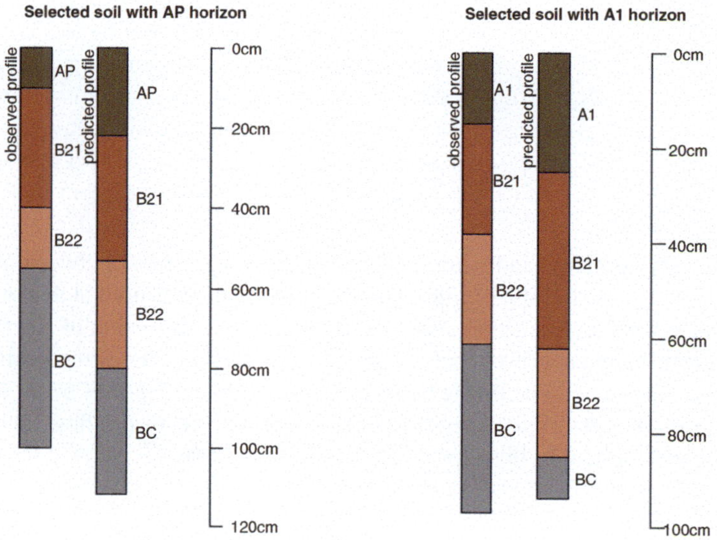

Fig. 9.3 Examples of observed soil profiles with associated predicted profiles from the two-stage horizon class and horizon depth model

```
p1u <- c(0, (0 + p1[1]), (0 + p1[1] + p1[2]), (0 + p1[1] + p1[2] + p1[3]))
p1l <- c(p1[1], (p1[1] + p1[2]), (p1[1] + p1[2] +p1[3]), (p1[1] +
p1[2] + p1[3] + p1[4]))
p2u <- c(0, (0 + p2[1]), (+p2[1] + p2[2]), (+p2[1] + p2[2] + p2[3]))
p2l <- c(p2[1], (p2[1] + p2[2]), (p2[1] + p2[2] + p2[3]), (p2[1] +
p2[2] + p2[3] +p2[4]))

# Upper depths
U1 <- c(p1u, p2u)
# Lower depths
L1 <- c(p1l, p2l)

# Soil profile names
S1 <- c("predicted profile", "predicted profile",
"predicted profile", "predicted profile", "observed profile",
"observed profile", "observed profile", "observed profile")

# Random soil colors selected to distinguish between horizons
hue <- c("10YR", "10R", "10R", "10R", "10YR", "10R", "10R", "10R")
val <- c(4, 5, 7, 6, 4, 5, 7, 6)
chr <- c(3, 8, 8, 1, 3, 8, 8, 1)

# Combine all the data
TT1 <- data.frame(S1, U1, L1, H1, hue, val, chr)
```

```
# Convert munsell colors to rgb
TT1$soil_color <- with(TT1, munsell2rgb(hue, val, chr))

# Upgrade to soil profile collection
depths(TT1) <- S1 ~ U1 + L1

# Plot
plot(TT1, name = "H1", colors = "soil_color")
title("Selected soil with AP horizon", cex.main = 0.75)
```

Soil is very complex, and while there is a general agreement between observed and associated predicted soil profiles, the power of the models used in this two-stage example has certainly been challenged. Recreating the arrangement of soil horizons together with maintenance of their depth properties is an interesting problem for pedometric studies and one that is likely to be pursued with vigor as better methods become available. The next section will briefly demonstrate a work flow for creating maps that are resultant of this type of modeling framework.

9.2 Spatial Application of the Two-Stage Soil Horizon Occurrence and Depth Model

We will recall from previous chapters the process for applying prediction models across a mapping extent. In the case of the two-stage model the mapping work flow if first creating the map of horizon presence/occurrence. Then we apply the horizon depth model. In order to ensure that the depth model is not applied to the areas where a particular soil horizon is predicted as being absent, those areas are masked out. Maps for the presence of the A1 and AP horizons and their respective depths are displayed in Fig. 9.4. The following scripts show the process of applying the two-stage model for the A1 horizon.

```
# Apply A1 horizon presence/absence model spatially Using
# the raster multi-core facility
beginCluster(4)
A1.class <- clusterR(s1, predict, args = list(mn2, type = "class"),
filename = "class_A1.tif", format = "GTiff", progress = "text",
overwrite = T)

# Apply A1 horizon depth model spatially Using the raster
# multi-core facility
A1.depth <- clusterR(s1, predict, args = list(qrf, index = 2),
filename = "rasterPreds/depth_A1.tif", format = "GTiff",
progress = "text", overwrite = T)
endCluster()

# Mask out areas where horizon is absent
A1.class[A1.class == 0] <- NA
```

```
mr <- mask(A1.depth, A1.class)
writeRaster(mr, filename = "depth_A1_mask.tif", format = "GTiff",
overwrite = T)
```

Figure 9.4 displays an interesting pattern whereby AP horizons occur where A1 horizons do not. This makes reasonable sense. The spatial pattern of the AP horizon coincides generally with the distribution of vineyards across the study area, where soils are often cultivated and consequently removing the presence of the A1 horizon. Increased depth of A1 horizons also appears to be the case too in lower lying and stream channel catchments of the study area.

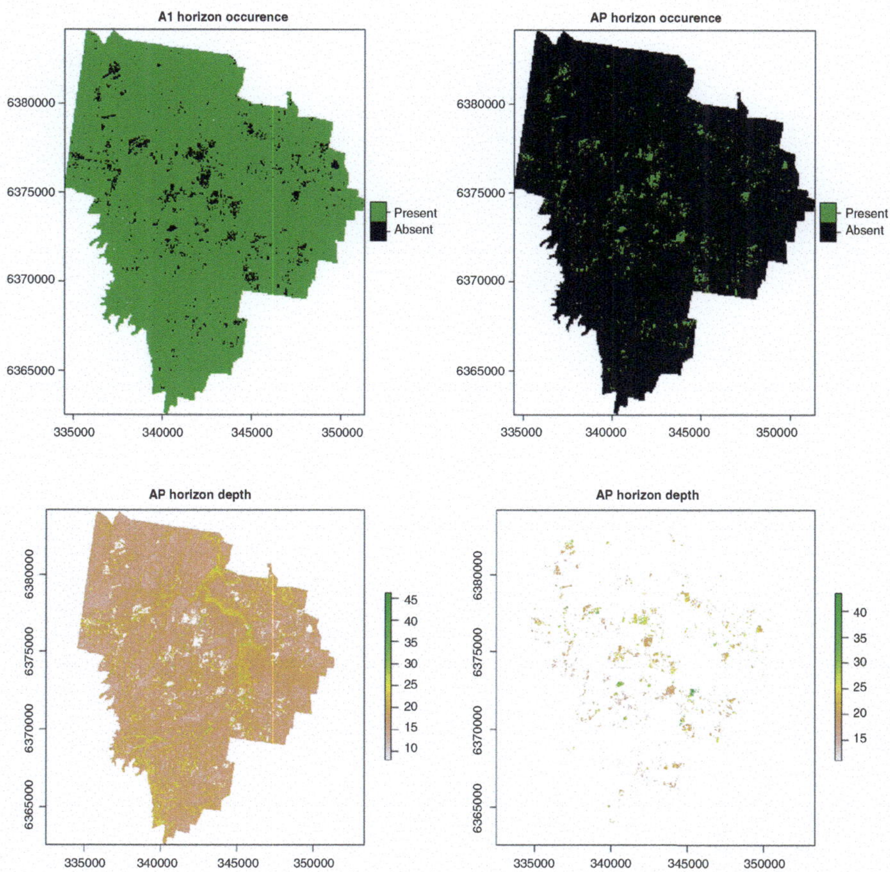

Fig. 9.4 Predicted occurrence of AP and A1 horizons, and their respective depths in the Lower Hunter Valley, NSW

References

Breiman L (2001) Random forests. Mach Learn 41:5–32

Gastaldi G, Minasny B, McBratney AB (2012) Mapping the occurrence and thickness of soil horizons. In: Minasny B, Malone BP, McBratney AB (eds) Digital soil assessments and beyond. Taylor & Francis, London, pp 145–148

Isbell RF (2002) The Australian soil classification, rev. edn. CSIRO Publishing, Collingwood

Malone BP, Kidd DB, Minasny B, McBratney AB (2015) Taking account of uncertainties in digital land suitability assessment. PeerJ 3:e1366

Meinshausen N (2006) Quantile regression forests. J Mach Learn Res 7:983–999

Sileshi G (2008) The excess-zero problem in soil animal count data and choice of appropriate models for statistical inference. Pedobiologia 52(1):1–17

Chapter 10
Digital Soil Assessments

Digital soil assessment goes beyond the goals of digital soil mapping. Digital soil assessment (DSA) can be defined (from McBratney et al. (2012)) as the translation of digital soil mapping outputs into decision making aids that are framed by the particular, contextual human-value system which addresses the question/s at hand. The concept of DSA was first framed by Carre et al. (2007) as a mechanism for assessing soil threats, assessing soil functions and for soil mechanistic simulations to assess risk based scenarios to complement policy development. Very simply DSA can be likened to the quantitative modeling of difficult-to-measure soil attributes. An obvious candidate application for DSA is land suitability evaluation for a specified land use type, which van Diepen et al. (1991) define as all methods to explain or predict the use potential of land. The first part of this chapter will cover a simple digital approach for performing this type of analysis. The second part of the chapter will explore a different form of digital assessment by way of identifying soil homologues (Mallavan et al. 2010).

10.1 A Simple Enterprise Suitability Example

Land evaluation in some sense has been in practice at least since the earliest known human civilizations. The shift to sedentary agriculture from nomadic lifestyles is at least indicative of a concerted effort of human investment to evaluate the potential and capacity of land and its soils to support some form of agriculture like cropping (Brevik and Hartemink 2010). In the modern times there is a well-documented history of land evaluation practice and programs throughout the world, many of which are described in Mueller et al. (2010). Much of the current thinking around land evaluation for agriculture are well documented within the land evaluation guidelines prepared by the Food and Agriculture Organization of the United Nations (FAO) in 1976 (FAO 1976). These guidelines have strongly influenced and continue

B.P. Malone et al., *Using R for Digital Soil Mapping*,
Progress in Soil Science, DOI 10.1007/978-3-319-44327-0_10

to guide land evaluation projects throughout the world. The FAO framework is a crop specific LSA system with a 5-class ranking of suitability (FAO Land Suitability Classes) from 1: Highly Suitable to 5: Permanently Not Suitable. Given a suite of biophysical information from a site, each attribute is evaluated against some expert-defined thresholds for each suitability class. The final evaluation of suitability for the site is the one in which is most limiting.

Digital soil mapping complementing land evaluation assessment is being more regularly observed. Examples (in Australia) include Kidd et al. (2012) in Tasmania and Harms et al. (2015) in Queensland. Perhaps an obvious reason is that one can derive with digital soil and climate modeling, very attribute specific mapping which can be targeted specifically to a particular agricultural land use or even to a specific enterprise (Kidd et al. 2012).

In this chapter, an example is given of a DSA where enterprise suitability is assessed. The specific example is a digital land suitability assessment (LSA) for hazelnuts across an area of northern Tasmania, Australia (Meander Valley) which has been previously described in Malone et al. (2015). For context, the digital soil assessment example has been one function of the Tasmanian Wealth from Water project for developing detailed land suitability assessments (20 specific agricultural enterprises) to support irrigated agricultural expansion across the state (Kidd et al. 2015, 2012). The project was commissioned for a couple of targeted areas, but has since been rolled out across the state (Kidd et al. 2015). Further general information about the project can be found at http://dpipwe.tas.gov.au/agriculture/investing-in-irrigation/enterprise-suitability-toolkit.

The example considered in this chapter is just to give an overview of how to perform a DSA in what could be considered as a relatively simple example. Using the most-limiting factor approach of land suitability assessment, the procedure requires a pixel-by-pixel assessment of a number of input variables which have been expertly defined as being important for the establishment and growth of hazelnuts. Malone et al. (2015) describes the digital mapping processes that went into creating the input variables for this example. The approach also assumes that the predicted maps of the input variables are also error free. Figure 10.1 shows an example of the input variable requirements and the suitability thresholds for hazelnuts. You will notice the biophysical variables include both soil and climatic variables, and the suitability classification has four levels of grading.

Probably the first thing to consider for enabling the DSA in this example is to codify the information in Fig. 10.1 into an `R` function. It would look like something similar to the following script.

```
# HAZELNUT SUITABILITY ASSESSMENT FUNCTION
hazelnutSuits <- function(samp.matrix) {
out.matrix <- matrix(NA, nrow = nrow(samp.matrix), ncol = 10)

# Chill Hours
out.matrix[which(samp.matrix[, 1] > 1200), 1] <- 1
out.matrix[which(samp.matrix[, 1] > 600 & samp.matrix[, 1] <= 1200), 1] <- 2
out.matrix[which(samp.matrix[, 1] <= 600), 1] <- 4
```

Crop	Soil Depth	Depth to sodic layer	pH of top 15cm (H20)	EC (top 15cm)	Texture (top 15cm-% clay))	Drainage	Stoniness (top 15cm)	Frost	Mean max monthly temp	rainfall	Chill hours
W	>50cm		6.5	<0.15dS/m	10-30%	Well, Moderately well	<10%<=2 (>200mm)	No days <-6 deg C in June,July or Aug– occurs 4/5 years	Mean Jan or Feb max temp -20-30°C	<50mm (mean march)	Chill hours 0-7°C (April-August inclusive):>1200
S	40-50cm		5.5-6.5	<0.15dS/m	30-50%	Imperfect	10-20% 3 (>200mm)	No days <-6 deg C in June,July or Aug– occurs 3/5 to 4/5 years	Mean Jan or Feb max temp –30-33°C & 18-20°C	<50mm (mean march)	Chill hours 0-7°C (April-August inclusive):600-1200
MS	30-40cm		6.5-7.1	<0.15dS/m	30-50%	Imperfect	10-20% 4 (>200mm)	No days <-6 deg C in June,July or Aug– occurs 2/5 to 3/5 years	Mean Jan or Feb max temp –33-35°C	<50mm (mean march)	Chill hours 0-7°C (April-August inclusive):600-1200
U	<30cm		<5.5 >7.1	>0.15dS/m	>50% or <10%	Poor,Very poor	>20%>= 4 (>200mm)	No days <-6 deg C in June,July or Aug– occurs <2/5 years	Mean Jan or Feb max temp –>35°C & <18°C	>50mm (mean march)	Chill hours 0-7°C (April-August inclusive):<600

Well suited (W): Land having no significant limitations to sustained applications of a given use, or only minor limitations that will not significantly reduce productivity or benefits and will not raise inputs above an acceptable level. Any risk of crop loss is inherently low or can be easily overcome with management practices that are easy and cheap to implement.

Suitable (S): Land having no limitations which are moderately severe for sustained applications of a given use; the limitations will reduce productivity or benefits and increase required inputs to the extent that the overall advantage to be gained from the use, although still attractive, will be appreciable inferior to that expected on Class S1 land. Risk of crop loss is moderately high or requires management practices that are dificult or costly to implement.

Marginally Suitable(MS): Land having no limitations which are which are severe for sustained application of a given use and will so reduce productivity or benefits,or increase required inputs, that this expenditure will be only marginally justified. Risk of crop loss may be high.

Unsuitable (U): Land which has qualities that appear to preclude sustained use of the kind under consideration

Fig. 10.1 Suitability parameters and thresholds for hazelnuts (Sourced from DPIPWE 2015)

```
# Clay content
out.matrix[which(samp.matrix[, 2] > 10 & samp.matrix[, 2] <= 30), 2] <- 1
out.matrix[which(samp.matrix[, 2] > 30 & samp.matrix[, 2] <= 50), 2] <- 2
out.matrix[which(samp.matrix[, 2] > 50 | samp.matrix[, 2] <= 10), 2] <- 4

# Soil Drainage
out.matrix[which(samp.matrix[, 3] > 3.5), 3] <- 1
out.matrix[which(samp.matrix[, 3] <= 3.5 & samp.matrix[, 3] > 2.5), 3] <- 2
out.matrix[which(samp.matrix[, 3] <= 2.5 & samp.matrix[, 3] > 1.5), 3] <- 3
out.matrix[which(samp.matrix[, 3] <= 1.5), 3] <- 1

# EC (transformed variable)
out.matrix[which(samp.matrix[, 4] <= 0.15), 4] <- 1
out.matrix[which(samp.matrix[, 4] > 0.15), 4] <- 4

# Frost
out.matrix[which(samp.matrix[, 10] == 0), 5] <- 1
out.matrix[which(samp.matrix[, 10] != 0 & samp.matrix[, 5] >= 80), 5] <- 1
out.matrix[which(samp.matrix[, 10] != 0 & samp.matrix[, 5] < 80 & samp.matrix
[, 5] >= 60), 5] <- 2
out.matrix[which(samp.matrix[, 10] != 0 & samp.matrix[, 5] < 60 & samp.matrix
[, 5] >= 40), 5] <- 3
out.matrix[which(samp.matrix[, 10] != 0 & samp.matrix[, 5] < 40), 5] <- 3

# pH
out.matrix[which(samp.matrix[, 6] <= 6.5 & samp.matrix[, 6] >= 5.5), 6] <- 1
out.matrix[which(samp.matrix[, 6] > 6.5 & samp.matrix[, 6] <= 7.1), 6] <- 3
out.matrix[which(samp.matrix[, 6] < 5.5 | samp.matrix[, 6] > 7.1), 6] <- 4

# rainfall
out.matrix[which(samp.matrix[, 7] <= 50), 7] <- 1
out.matrix[which(samp.matrix[, 7] > 50), 7] <- 4
```

```
# soil depth
out.matrix[which(samp.matrix[, 13] == 0), 8] <- 1
out.matrix[which(samp.matrix[, 13] != 0 & samp.matrix[, 8] > 50), 8] <- 1
out.matrix[which(samp.matrix[, 13] != 0 & samp.matrix[, 8] <= 50 & samp.matrix
[, 8] > 40), 8] <- 2
out.matrix[which(samp.matrix[, 13] != 0 & samp.matrix[, 8] <= 40 & samp.matrix
[, 8] > 30), 8] <- 3
out.matrix[which(samp.matrix[, 13] != 0 & samp.matrix[, 8] <= 30), 8] <- 4

# temperature
out.matrix[which(samp.matrix[, 9] > 20 & samp.matrix[, 9] <= 30), 9] <- 1
out.matrix[which(samp.matrix[, 9] > 30 & samp.matrix[, 9] <= 33 | samp.matrix
[, 9] <= 20 & samp.matrix[, 9] > 18), 9] <- 2
out.matrix[which(samp.matrix[, 9] > 33 & samp.matrix[, 9] <= 35), 9] <- 3
out.matrix[which(samp.matrix[, 9] > 35 | samp.matrix[, 9] <= 18), 9] <- 4

# rocks
out.matrix[which(samp.matrix[, 11] == 0), 10] <- 1
out.matrix[which(samp.matrix[, 11] != 0 & samp.matrix[, 12] <= 2), 10] <- 1
out.matrix[which(samp.matrix[, 11] != 0 & samp.matrix[, 12] == 3), 10] <- 2
out.matrix[which(samp.matrix[, 11] != 0 & samp.matrix[, 12] == 4), 10] <- 3
out.matrix[which(samp.matrix[, 11] != 0 & samp.matrix[, 12] > 4), 10] <- 4
return(out.matrix)
}
```

Essentially the function takes in a matrix of (n) number of rows by 13 columns. The number of columns is fixed as each coincides with one of the biophysical variables. There are some variables where there is relevant information contained in two columns. For example, columns assigned to soil depth information are contained in columns 8 and 13. The reason is such that soil depth is modeled via a two-stage process (as is exemplified in the chapter regarding two-stage modeling for DSM), such that a model is used to predict the presence/absence of a certain condition—which for soil depth is whether lithic contact is achieved less than 1.5 m from the soil surface—followed by a secondary model that predicts soil depth (where soil is expected to be less than 1.5m). Subsequently the secondary model is only invoked if there is a positive condition found for the first model. As such the above `R` function provides a good example of the use of sub-setting and applying conditional queries within `R`. The function `hazelnutSuits` will return a new matrix with *n* rows and 10 columns. The entries for each column and row will be the suitability assessment for each biophysical variable. It is then just a matter of determining what the maximum value is on each row in order to give an overall suitability valuation (as this is the most limiting factor approach). This can be achieved using the `rowMaxs` function from the `matrixStats` package.

```
rowMaxs(output from hazelnutSuits function)
```

Following is a workflow for implementing the `hazelnutSuits` function, with special application of it in a spatial mapping context.

10.1.1 Mapping Example of Digital Land Suitability Assessment

Assuming there are digital soil and climate maps already created for use in the hazelnut land suitability assessment, it is relatively straightforward to run the LSA. Keep in mind that the creation of the biophysical variable maps were created via a number of means which included continuous attribute modeling, binomial and ordinal logistic regression, and a combination of both i.e. through the two-stage mapping process. So let's get some sense of the LSA input variables.

```
# LSA input variables
library(raster)
names(lsa.variables)

##  [1] "X1_chill_HAZEL_FINAL_meander"
##  [2] "X2_clay_FINAL_meander"
##  [3] "X3_drain_FINAL_meander"
##  [4] "X4_EC_cubist_meander"
##  [5] "X5_Frost_HAZEL_FINAL_meander"
##  [6] "X6_pH_FINAL_meander"
##  [7] "X7_rain_HAZEL_FINAL_meander"
##  [8] "X8_soilDepth_FINAL_meander"
##  [9] "X9_temp_HAZEL_FINAL_meander"
## [10] "X5_Frost_HAZEL_binaryClass_meander"
## [11] "X10_rocks_binaryClass_meander"
## [12] "X11_rocks_ordinalClass_meander"
## [13] "X8_soilDepth_binaryClass_meander"

class(lsa.variables)

## [1] "RasterStack"
## attr(,"package")
## [1] "raster"

# Raster stack dimensions
dim(lsa.variables)

## [1] 1081 1685   13

# Raster resolution
res(lsa.variables)

## [1] 30 30
```

So there are 13 rasters of data, which you will note coincide with the inputs required for the `hazelnutSuits` function. Now all that is required is to go pixel by pixel and apply the LSA function. In `R` the implementation may take the following form:

```
# Retrieve values of from all rasters for a given row
#  Here it is the 1st row of the raster
cov.Frame <- getValues(lsa.variables, 1)
nrow(cov.Frame)

## [1] 1685

# Remove the pixels where there is NA or no data present
sub.frame <- cov.Frame[which(complete.cases(cov.Frame)), ]
nrow(sub.frame)

## [1] 27

names(sub.frame)

## NULL

# run hazelnutSuits function
hazel.lsa <- hazelnutSuits(sub.frame)

# Assess for suitability
library(matrixStats)
rowMaxs(hazel.lsa)

##  [1] 2 2 2 2 2 2 2 2 2 2 2 2 2 2 2 2 2 2 2 2 4 4 4 4 4 4 4
```

The above script has just performed the hazelnut LSA upon the entire first row of the input variable rasters. There were only 27 pixels where the full suite of data was available in this case. To do the LSA for the entire mapping extent we could effectively run the script above for each row of the input rasters. Naturally, this would take an age to do manually, so it might be more appropriate to use the script above inside a `for` loop where the row index changes for each loop. Alternatively, the custom `hazelnutSuits` function could be used as an input argument for the `raster` package `calc` function, or better still using the `clusterR` function if there is a need to do the LSA in parallel mode across multiple compute nodes. Given the available options, we will demonstrate the mapping process using the looping approach. While it may be computationally slower, it imprints the concept of applying the LSA spatially with greater clarity.

From the above script and resulting output, we essentially returned a vector of integers. We effectively lost all links to the fact that the inputs and resulting outputs are spatial data. Subsequently there is a need to link the suitability assessment back to the mapping. Fortunately we know the column positions of the instances where there was the full suite of input data. So once we have set up a raster object to which data can be written to (which has the same raster properties of the input data), it is just a matter of placing the LSA outputs into the row and column positions as those of the input data. A full example is given below, but for the present purpose, the LSA output data placement into a raster would look something like the following:

```
# A one column matrix with number of rows equal to number of
# columns in raser inputs
a.matrix <- matrix(NA, nrow = nrow(cov.Frame), ncol = 1)
```

```
# Place LSA outputs into correct row positions
a.matrix[which(complete.cases(cov.Frame)), 1] <- rowMaxs(hazel.lsa)

# Write LSA outputs to raster object (1st row)
LSA.raster <- writeValues(LSA.raster, a.matrix[, 1], 1)
```

Naturally the above few lines of script would also be embedded into the looping process as described above. Below is an example of putting all these operations together to ultimately produce a map of the suitability assessment. To add a slight layer of complexity, we may also want to produce maps of the suitability assessment for each of the input variables. This helps in determining which factors are causing the greatest limitations and where they occur. First, we need to create a number of rasters to which we can write the outputs of the LSA to.

```
# Create a suite of rasters of same raster properties
# is LSA input variables
# Overall suitability classification
LSA.raster <- raster(lsa.variables[[1]])

# Write outputs of LSA directly to file
LSA.raster <- writeStart(LSA.raster,
filename = "meander_MLcat.tif", format = "GTiff",
dataType = "INT1S", overwrite = TRUE)

# Individual LSA input suitability rasters
mlf1 <- raster(lsa.variables[[1]])
mlf2 <- raster(lsa.variables[[1]])
mlf3 <- raster(lsa.variables[[1]])
mlf4 <- raster(lsa.variables[[1]])
mlf5 <- raster(lsa.variables[[1]])
mlf6 <- raster(lsa.variables[[1]])
mlf7 <- raster(lsa.variables[[1]])
mlf8 <- raster(lsa.variables[[1]])
mlf9 <- raster(lsa.variables[[1]])
mlf10 <- raster(lsa.variables[[1]])

# Also write LSA outputs directly to file.
mlf1 <- writeStart(mlf1, filename = "meander_Haz_mlf1_CAT.tif",
format = "GTiff", dataType = "INT1S", overwrite = TRUE)
mlf2 <- writeStart(mlf2, filename = "meander_Haz_mlf2_CAT.tif",
format = "GTiff",dataType = "INT1S", overwrite = TRUE)
mlf3 <- writeStart(mlf3, filename = "meander_Haz_mlf3_CAT.tif",
format = "GTiff",dataType = "INT1S", overwrite = TRUE)
mlf4 <- writeStart(mlf4, filename = "meander_Haz_mlf4_CAT.tif",
format = "GTiff",dataType = "INT1S", overwrite = TRUE)
mlf5 <- writeStart(mlf5, filename = "meander_Haz_mlf5_CAT.tif",
format = "GTiff", dataType = "INT1S", overwrite = TRUE)
mlf6 <- writeStart(mlf6, filename = "meander_Haz_mlf6_CAT.tif",
format = "GTiff", dataType = "INT1S", overwrite = TRUE)
```

```
mlf7 <- writeStart(mlf7, filename = "meander_Haz_mlf7_CAT.tif",
format = "GTiff", dataType = "INT1S", overwrite = TRUE)
mlf8 <- writeStart(mlf8, filename = "meander_Haz_mlf8_CAT.tif",
format = "GTiff", dataType = "INT1S", overwrite = TRUE)
mlf9 <- writeStart(mlf9, filename = "meander_Haz_mlf9_CAT.tif",
format = "GTiff", dataType = "INT1S", overwrite = TRUE)
mlf10 <- writeStart(mlf10, filename = "meander_Haz_mlf10_CAT.tif",
format = "GTiff", dataType = "INT1S", overwrite = TRUE)
```

Now we can implement the `for` loop procedure and do the LSA for the entire mapping extent.

```
# Run the suitability model: Open loop:for each row of each input
raster get
# raster values
for (i in 1:dim(LSA.raster)[1]) {
    cov.Frame <- getValues(lsa.variables, i)
    # get the complete cases
    sub.frame <- cov.Frame[which(complete.cases(cov.Frame)), ]

    # Run hazelnut LSA function
    t1 <- hazelnutSuits(sub.frame)

    # Save results to raster
    a.matrix <- matrix(NA, nrow = nrow(cov.Frame), ncol = 1)
    a.matrix[which(complete.cases(cov.Frame)), 1] <- rowMaxs(t1)
    LSA.raster <- writeValues(LSA.raster, a.matrix[, 1], i)

    # Also save the single input variable assessment outputs
    mlf.out <- matrix(NA, nrow = nrow(cov.Frame), ncol = 10)
    mlf.out[which(complete.cases(cov.Frame)), 1] <- t1[, 1]
    mlf.out[which(complete.cases(cov.Frame)), 2] <- t1[, 2]
    mlf.out[which(complete.cases(cov.Frame)), 3] <- t1[, 3]
    mlf.out[which(complete.cases(cov.Frame)), 4] <- t1[, 4]
    mlf.out[which(complete.cases(cov.Frame)), 5] <- t1[, 5]
    mlf.out[which(complete.cases(cov.Frame)), 6] <- t1[, 6]
    mlf.out[which(complete.cases(cov.Frame)), 7] <- t1[, 7]
    mlf.out[which(complete.cases(cov.Frame)), 8] <- t1[, 8]
    mlf.out[which(complete.cases(cov.Frame)), 9] <- t1[, 9]
    mlf.out[which(complete.cases(cov.Frame)), 10] <- t1[, 10]
    mlf1 <- writeValues(mlf1, mlf.out[, 1], i)
    mlf2 <- writeValues(mlf2, mlf.out[, 2], i)
    mlf3 <- writeValues(mlf3, mlf.out[, 3], i)
    mlf4 <- writeValues(mlf4, mlf.out[, 4], i)
    mlf5 <- writeValues(mlf5, mlf.out[, 5], i)
    mlf6 <- writeValues(mlf6, mlf.out[, 6], i)
    mlf7 <- writeValues(mlf7, mlf.out[, 7], i)
    mlf8 <- writeValues(mlf8, mlf.out[, 8], i)
    mlf9 <- writeValues(mlf9, mlf.out[, 9], i)
    mlf10 <- writeValues(mlf10, mlf.out[, 10], i)
    print((dim(LSA.raster)[1]) - i)
}  #END OF LOOP
```

```
# complete writing rasters to file
LSA.raster <- writeStop(LSA.raster)
mlf1 <- writeStop(mlf1)
mlf2 <- writeStop(mlf2)
mlf3 <- writeStop(mlf3)
mlf4 <- writeStop(mlf4)
mlf5 <- writeStop(mlf5)
mlf6 <- writeStop(mlf6)
mlf7 <- writeStop(mlf7)
mlf8 <- writeStop(mlf8)
mlf9 <- writeStop(mlf9)
mlf10 <- writeStop(mlf10)
```

As you may encounter, the above script can take quite a while to complete, but ultimately you should be able to produce a number of mapping products. Figure 10.2 shows the map of the overall suitability classification and the script to produce it is below.

```
library(rasterVis)

LSA.raster <- as.factor(LSA.raster)
rat <- levels(LSA.raster)[[1]]
rat[["suitability"]] <- c("Well Suited", "Suited",
"Moderately Suited", "Unsuited") levels(LSA.raster) <- rat

# plot
area_colors <- c("#FFFF00", "#1D0BE0", "#1CEB15", "#C91601")
levelplot(LSA.raster, col.regions = area_colors, xlab = "",
   ylab = "")
```

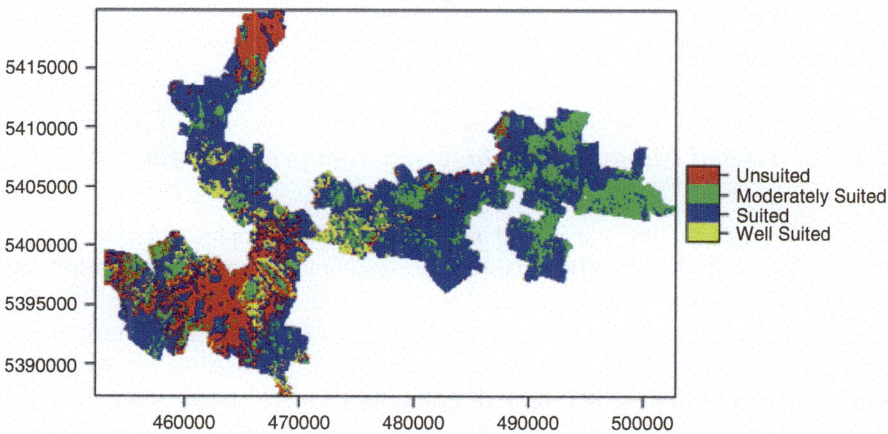

Fig. 10.2 Digital suitability assessment for hazelnuts across the Meander Valley, Tasmania (assuming all LSA input variables are error free)

Similarly the above plotting procedure can be repeated to look at single input variable limitations too.

The approaches detailed in this chapter are described in greater detail in Malone et al. (2015) within the context of taking account of uncertainties within LSA. Taking account of the input variable uncertainties adds an additional level of complexity to what was achieved above, but is an important consideration nonetheless, as the resulting outputs can be assessed for reliability in an objective manner. However, that particular workflow for LSA is not covered in this chapter as it is only meant to provide a general perspective and relatively simple example of a real world digital soil assessment.

10.2 Homosoil: A Procedure for Identifying Areas with Similar Soil Forming Factors

In many places in the world, soil information is difficult to obtain or even be non-existent. When no detailed maps or soil observations are available in a region of interest, we have to interpolate or extrapolate from other parts of the world. When dealing with global modeling at a coarse resolution, we can interpolate or extrapolate soil observations available from other similar areas (that are geographically close) or by using spatial interpolation or a spatial soil prediction function.

Homosoil, is a concept proposed by Mallavan et al. (2010) which assumes the homology of predictive soil-forming factors between a reference area and the region of interest. These include: climate, parent materials, and physiography of the area. We created the `homosoil` function to illustrate the concept. It is relatively simple whereby, given any location (latitude, longitude) in the world, the function will determine other areas in the world that share similar climate, lithology and topography. Shortly we will unpack the function into its elemental components. First we will describe the data and how to measure similarity between sites.

10.2.1 Global Climate, Lithology and Topography Data

The basis is a global $0.5° \times 0.5°$ grid data of climate, topography, and lithology.

For climate, this consists of variables representing long-term mean monthly and seasonal temperature, rainfall, solar radiation and evapotranspiration data. We also use the DEM representing topography, and lithology, which gives broad information on the parent material. The climate data come from the ERA-40 reanalysis and Climate Research Unit (CRU) dataset. More details on the datasets are available on the website http://www.ipcc-data.org/obs/get_30yr_means.html. For each of the 4 climatic variables (rainfall, temperature, solar radiation and evapotranspiration), we calculated 13 indicators: annual mean, mean for the driest month, mean at the wettest month, annual range, driest quarter mean, wettest quarter mean, coldest quarter mean, hottest quarter mean, lowest ET quarter mean, highest ET quarter mean, darkest quarter mean, lightest quarter mean, and seasonality. From this analysis and including the acquired data, 52 global climatic variables were composed.

The DEM is from the Hydro1k dataset supplied from the USGS (https://lta.cr.usgs.gov/HYDRO1K), which includes the mean elevation, slope, and compound topgraphic index (CTI).

The lithology is from a global digital map (Durr et al. 2005) with seven values which represent the different broad groups of parent materials. The lithology classes are: non- or semi-consolidated sediments, mixed consolidated sediments, silic-clastic sediments, acid volcanic rocks, basic volcanic rocks, complex of metamorphic and igneous rocks, and complex lithology.

This global data is available as a `data.frame` from the `ithir` package.

```
library(ithir)
data(homosoil_globeDat)
```

10.2.2 Estimation of Similarity

The climatic and topographic similarity between two points of the grid is calculated by a Gower similarity measure (Gower 1971). The Gower distance measures the similarity S_{ij} between sites i and j, each with p number of variables, standardized by the range of each variable:

$$S_{ij} = \frac{1}{p} \sum_{k=1}^{p} \left(1 - \frac{| x_{ik} - x_{jk} |}{range(k)} \right) \tag{10.1}$$

Where p denotes the number of climatic variables, $| x_{ik} - x_{jk} |$ represents the absolute difference of climate variables between site i and j. The similarity index has a value between 0 and 1 and is applicable for continuous variables. For categorical variables such as those for lithology, we simply just want to match category for category between sites i and j.

Considering the scale and the resolution of this study and the available global data (0.5° × 0.5°), the climatic factor is probably the most important and reliable soil forming factor. This is inspired by the study of Bui et al. (2006) who showed at continental extent, the state factors of soil formation form a hierarchy of interacting variables, with climate being the most important factor, and different climatic variables dominate in different regions. This is not to say that climate to be the most important factor at all scales. Their results also show that lithology is almost equally important in defining broad scale spatial patterns of soil properties, and shorter-range variability in soil properties appears to be driven more by terrain variables.

In the homosoil function, we first is to identify homoclimes around the world, and within the homoclime, we find areas with similar lithology (homolith), within which we then find similar topography (homotop). For climate and topography, we arbitrarily select the top x% similarity index as areas of homologue. In the following example we select the top 15 %. So let's look at the internals of the function before running an example.

10.2.3 *The* homosoil *Function*

Essentially from below we are creating the homosoil function. This function takes three inputs: grid.data, which is the global environmental dataset, and recipient.lon and recipient.lat, which correspond the coordinates of the sites to which we want to find soil homologues. For brevity we will call this the recipient site. Inside the homosoil function, we first encounter another function which is an encoding of the Gower's similarity measure as defined previously. This is followed by a number of indexation steps (to make the following steps clearer to implement), where we explicitly make groupings of the data, for example the object grid.climate is composed of all the global climate information from the grid.data object. Finally make the object world.grid which is a data.frame for putting outputs of the function into. Ultimately this object will get returned at the end of the function execution.

```
homosoil <- function (grid_data,recipient.lon,recipient.lat) {
#Gower's similarity function
gower <- function(c1, c2, r) 1-(abs(c1-c2)/r)

#index global data
grid.lon <-grid_data[,1] #longitude
grid.lat <-grid_data[,2] #latitude
grid.climate <-grid_data[,3:54]  #climate data
grid.lith <-grid_data[,58] #lithology
grid.topo <-grid_data[,55:57] #topography

#data frame to put outputs of homosoil function
world.grid<- data.frame(grid_data[, c("X", "Y")],
fid = seq(1, nrow(grid_data), 1), homologue = 0, homoclim=0,
homolith=0, homotop=0)
```

We then want to find which global grid point is the closest to the recipient site based on the Euclidean distance of the coordinates. We then want to extract the climate, lithological, and topographical data that is recorded for the nearest grid point.

```
# find the closest recipient point
dist = sqrt((recipient.lat - grid.lat)^2 + (recipient.lon
  - grid.lon)^2)
imin = which.min(dist)
```

```
# climate, lithology and topography for recipient site
recipient.climate <- grid.climate[imin, ]
recipient.lith <- grid.lith[imin]
recipient.topo <- grid.topo[imin, ]
```

Starting with climate, we want to estimate Gower's similarity measure. Firstly we estimate the range of values for each variable. Note the use of the `apply` function, which facilitates an efficient way to estimate ranges for each variable. Then using our `gower` function, we perform the similarity calculation. This is done for each variable. The `mapply` function allows us to do this without the need to use a `for` loop. We then take the mean of these values, to which corresponds to the Gower's similarity measure and which is saved to the object `Sr`.

```
# range of climate variables
rv <- apply(grid.climate, 2, range)
rr <- rv[2, ] - rv[1, ]

# calculate similarity to all variables in the grid
S <- (mapply(gower, c1 = grid.climate, c2 = recipient.climate,
   r = rr))

Sr <- apply(S, 1, mean)  # take the average
```

We can then determine which grid points are most similar to the recipient site. Here we use an arbitrarily selected cutoff of 0.85 which corresponds to the top 15 % of grid data similar to the recipient site. Lastly, we save the results of the homocline analysis to the `world.grid` object.

```
# row index for homoclime with top X% similarity.
iclim = which(Sr >= quantile(Sr, 0.85), arr.ind = TRUE)

# save homocline result
world.grid$homologue[iclim] <- 1
world.grid$homoclim[iclim] <- 1
```

Now we want to find within the areas we have defined as homocline, areas that are homolith. We simply want to find the lithology match between the recipient site and global lithology. We can do this for the entire globe, and then index those sites that also correspond to homoclines. Again, we save the results of the homolith analysis to the `world.grid` object.

```
# find within homoclime, areas with homolith
ilith = which(grid.lith == recipient.lith, arr.ind = TRUE)
#global comparison

# homolith in areas of homocline
clim.match <- which(world.grid$homologue == 1)
climlith.match <- clim.match[clim.match %in% ilith]
```

```
# save homolith result
world.grid$homologue[climlith.match] <- 2
world.grid$homolith[climlith.match] <- 1
```

Now we want to find within the areas we have defined as homolith, areas that are homotop. This analysis can be initiated by doing estimating the Gower's similarity measure for the whole globe. These steps below are just the same as before for the climate date, except now we are using the topographic data. Again we are also using the arbitrarily selected threshold value of 15 %.

```
# range of topographic variables
rv <- apply(grid.topo, 2, range)
rt <- rv[2, ] - rv[1, ]

# calculate similarity of topographic variables
Sa <- (mapply(gower, c1 = grid.topo, c2 = recipient.topo, r = rt))
St <- apply(Sa, 1, mean)  # take the average

# row index for homotop
itopo = which(St >= quantile(St, 0.85), arr.ind = TRUE)
```

Now we want to determine those areas that are homotop within the areas that are homolith.

```
top.match <- which(world.grid$homologue == 2)
lithtop.match <- top.match[top.match %in% itopo]

# save homotop result
world.grid$homologue[lithtop.match] <- 3
world.grid$homotop[lithtop.match] <- 1
```

That more-or-less completes the homosoil analysis. The last few tasks are to create a raster object of the soil homologues.

```
# homologue raster object
r1 <- rasterFromXYZ(world.grid[, c(1, 2, 4)])
r1 <- as.factor(r1)
rat <- levels(r1)[[1]]
rat[["homologue"]] <- c("", "homocline", "homolith", "homotop")
levels(r1) <- rat
```

Followed by directing the `homosoil` function to save the relevant outputs which here are the `world.grid` object and the raster object of the soil homologues. Then finally we close the function.

```
retval <- list(r1, world.grid)
return(retval)}
```

10.2.4 Example of Finding Soil Homologues

With the homosoil function now established, let's put it to use. The coordinates below correspond to a location in Jakarta in Indonesia.

```
recipient.lat = -(6 + 10/60)
recipient.lon = 106 + 49/60
```

Now we run the homosoil function.

```
result <- homosoil(grid_data = homosoil_globeDat,
recipient.lon = recipient.lon, recipient.lat = recipient.lat)
```

Then we plot the result. Here we want to use the map object that was created inside the homosoil function. We also specify colors to correspond the non-homologue areas, homoclines, homoliths, and homotops. Because of the hierarchical nature of the homosoil analysis, essentially the homotops are the soil homologues to the recipient site (Fig. 10.3).

```
# plot
area_colors <- c("#EFEFEF", "#666666", "#FFDAD4", "#FF0000")
levelplot(result[[1]], col.regions = area_colors,
xlab = "", ylab = "") + layer(sp.points(dats,
    col = "green", pch = 20, cex = 2))
```

Using the other object that is returned from the homosoil function (which was the world.grid data frame used to put the analysis outputs into) we can also map out the homologues individually, for example, we may just want to map the homoclines. That you can work out to do in your own time.

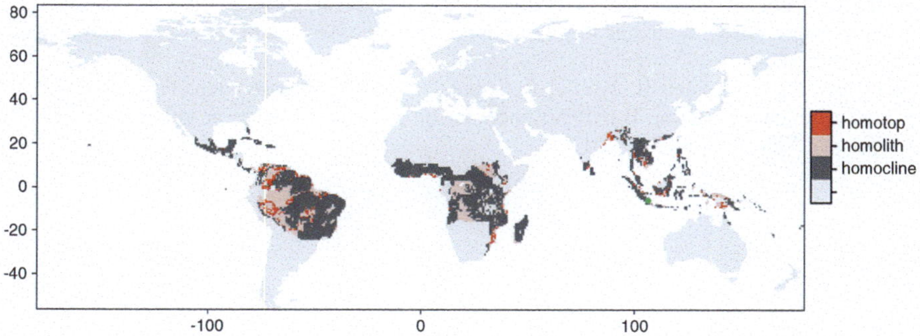

Fig. 10.3 Soil homologues to an area of Jakarta, Indonesia (*green dot* on map)

References

Brevik EC, Hartemink AE (2010) Early soil knowledge and the birth and development of soil science. Catena 83(1):23–33

Bui EN, Henderson BL, Viergever K (2006) Knowledge discovery from models of soil properties developed through data mining. Ecol Model 191:431–446

Carre F, McBratney AB, Mayr T, Montanarella L (2007) Digital soil assessments: beyond DSM. Geoderma 142(1–2):69–79

DPIPWE (2015) Enterprise suitability toolkit [online] dpipwe.tas.gov.au

Durr HH, Meybeck M, Durr SH (2005) Lithologic composition of the Earths continental surfaces derived from a new digital map emphasizing riverine material transfer. Glob Biogeochem Cycles 19:GB4S10

FAO (1976) A framework for land evaluation. Soils bulletin, vol 32. Food and Agriculture Organisation of the United Nations, Rome

Gower JC (1971) A general coefficient of similarity and some of its properties. Biometrics 27:857–871

Harms B, Brough D, Philip S, Bartley R, Clifford D, Thomas M, Willis R, Gregory L (2015) Digital soil assessment for regional agricultural land evaluation. Global Food Secur 5:25–36. Special Section on 3rd.

Kidd DB, Webb MA, Malone AB, Minasny B, McBratney AB (2015) Digital soil assessment of agricultural suitability, versatility and capital in tasmania, Australia. Geoderma Reg 6:7–21

Kidd DB, Webb MA, Grose CJ, Moreton RM, Malone BP, McBratney AB, Minasny B, Viscarra-Rossel R, Sparrow LA, Smith R (2012) Digital soil assessment: guiding irrigation expansion in Tasmania, Australia. In: Minasny B, Malone BP, McBratney AB (eds) Digital soil assessment and beyond. CRC, Boca Raton, pp 3–9

Mallavan BP, Minansy B, McBratney AB (2010) Homosoil: a methodology for quantitative extrapolation of soil information across the globe. In: Boettinger JL, Howell DW, Moore AC, Hartemink AE, Kienast-Brown S (eds) Digital soil mapping: bridging research, environmental application, and operation. Springer, New York, pp 137Ű149

Malone BP, Kidd DB, Minasny B, McBratney AB (2015) Taking account of uncertainties in digital land suitability assessment. PeerJ 3:e1366

McBratney AB, Minasny B, Wheeler I, Malone BP, Linden DVD (2012) Frameworks for digital soil assessment. In: Minasny B, Malone BP, McBratney AB (eds) Digital soil assessments and beyond. CRC Press, London, pp 9–15

Mueller L, Schindler U, Mirschel W, Shepherd T, Ball B, Helming K, Rogasik J, Eulenstein F, Wiggering H (2010) Assessing the productivity function of soils. A review. Agron Sustain Dev 30(3):601–614

van Diepen C, van Keulen H, Wolf J, Berkhout J (1991) Land evaluation: From intuition to quantification. In: Stewart B (ed) Advances in soil science. Advances in soil science, vol 15. Springer, New York, pp 139–204

Index

B.P. Malone et al., *Using R for Digital Soil Mapping*,
Progress in Soil Science, DOI 10.1007/978-3-319-44327-0